Climate Change in the Himalayas

G.B. Pant • P. Pradeep Kumar
Jayashree V. Revadekar • Narendra Singh

Climate Change in the Himalayas

Springer

G.B. Pant
Department of Atmospheric and Space Sciences
Savitribai Phule Pune University
Pune, Maharashtra, India

P. Pradeep Kumar
Department of Atmospheric and Space Sciences
Savitribai Phule Pune University
Pune, Maharashtra, India

Jayashree V. Revadekar
Centre for Climate Change Research
Indian Institute of Tropical Meteorology
Pune, Maharashtra, India

Narendra Singh
Atmospheric Sciences Group
Aryabhatta Research Institute of Observational Sciences
Nainital, Uttarakhand, India

ISBN 978-3-319-61653-7 ISBN 978-3-319-61654-4 (eBook)
DOI 10.1007/978-3-319-61654-4

Library of Congress Control Number: 2017944202

Printed on acid-free paper

This Springer imprint is published by Springer Nature
The registered company is Springer International Publishing AG
The registered company address is: Gewerbestrasse 11, 6330 Cham, Switzerland

Preface

Changes in the natural environment of the planet Earth, including its climate, have fascinated mankind all through the ages. The modern scientific understanding of natural processes that govern the functioning of the Earth system has provided sufficient knowledge to understand its present climate, including its evolution and future changes. The developments of the last few centuries have culminated into a system of monitoring and generation of global data sets, and enhancement in the basic theoretical understanding of complex processes involved in the science of Earth's climate. As a result of these efforts many holistic earth system models have evolved during the last few decades. These models have proven to be successful in generating plausible scenarios of future climates and highlighting the role of human interference in climate change. Realizing the role of anthropogenic influences on climate change, the last few decades have resulted in unprecedented activity and interest in the phenomenon of climate change in scientific, social, and political circles. This understanding and awareness has brought the issues of climate change impacts, mitigation, adaptability, and remedial measures at the forefront of all developmental activities with a focus on conserving the environment and the sustainability of natural resources. In spite of these accelerated developments, the major stakeholders with serious concern for their future as well as that of the generations ahead lack basic information and understanding on the subject to grasp and sensibly react to the issues in a broader perspective.

An academic career of almost half a century in the field of Earth system science has provided me an opportunity to be closer to many scientific ideas, deliberations, discussions, and differences of perceptions and goals at global and regional levels on climate change and other allied issues. It is, therefore, natural that the idea of writing a book as a basic source material on the subject suitable for a wide spectrum of scientifically oriented readers has always been in my mind. The opportunity to write this book acquired its renewed vigor after I joined the faculty at the Department of Atmospheric and Space Sciences at the Savitribai Phule Pune University. This has renewed my contacts with interested colleagues and students within and outside the university, and has created a sense of enthusiasm and confidence to take up this task as a joint effort. My coauthors, Dr. Pradeep Kumar, Dr. J.V. Revadekar, and

Dr. Narendra Singh have not only contributed in the subjects of their interest but also in every stage of the manuscript preparation. I may like to emphasize that in every sense this material is a joint contribution by all of us and this endeavor has enriched our joint research and teaching experiences.

Any writing in the subject with an intended wide spectrum of readership will not be able to fulfill its purpose unless it contains specific examples from a climate--sensitive target area. The authors have selected the region of Western Himalaya, possessing a unique weather system, large forest cover, extensive mountain ranges with rivers, glaciers, and snow cover as well as a highly fragile ecosystem for detailed discussion and illustrations. Western and Central Himalaya, the highest and largest mountain ranges of the world, are the major source of fresh water to many countries of the South Asian region. The weather and climate of Himalaya plays an important role in the atmospheric general circulation systems which affect the large population living in the region and its biodiversity and abundance of flora and fauna. Himalaya plays an important role in the establishment and sustainability of large--scale monsoon systems over South Asia. The book also very briefly discusses the monsoonal climate of the Indian region to illustrate the relevant features of regional climate with bearings on Himalayan weather and climate. The mountain ecosystem over the region is very delicate and highly susceptible to even minor changes in their complex environmental parameters, with a high degree of dependence on changes in global and local climate factors. Many scientific studies and reports, particularly the IPCC reports on mountain glaciers, have highlighted the irreversible nature of the impact of climate change on the mountain regions especially the melting of snow and an accelerated pace of retreating glaciers. The subject of Himalayan climate is very timely and of wider interest. It is expected that the topic will find general acceptance among climate scientists, meteorologists, water resource scientists, and wide a spectrum of social scientists and policy makers.

To fully incorporate certain recent studies and make the subject material more useful for understanding climate change impacts on a smaller region, we have downscaled our discussion to a subregion designated as Central Himalaya. Due to the limitations of specific long period station data to represent an inhomogeneous terrain, we depended on the data from few stations as well as the grid point data from global analysis sets for illustrations. The major part of Central Himalaya, mainly represented by the state of Uttarakhand, is considered and used in the discussion of climate change impacts, remedial measures, and adaptation policies. These sections are included primarily to provide some guidelines to policy makers for due consideration of climate change impact as one of the factors in their decision-making process. These sections specifically deal with core sectors such as water, agriculture, glaciers, forests, biodiversity, and natural disasters. In the process of writing on many facets of the subject, the information has been scrutinized from research papers, reports, documents, and electronic media on public domain.

We sincerely wish that the reader interested in the subject will find the contents of the book useful and informative. This is just a beginning as far as the Himalayas are concerned. The next generation of climate scientists will enlarge the climate change information base with an advantage of long period high resolution data over the Himalayas at their command, and further efforts will continue.

Pune, Maharashtra, India
January, 2017

Govind Ballabh Pant

Acknowledgements

Writing of a Book on the subject encompassing Himalayan dimensions and complex subject of climate change turned out to be much more of a challenging task than we initially perceived it to be. Complexity of the science of climate change coupled with limited and restricted data on many parameters, over the region selected for the study, increased the challenge. A group of four authors contributing on their expertise relevant to the subject in simple language to a broad spectrum of readers with varying interest has its own problems of consistency and balance, which we believe to have been reasonably achieved. Support and guidance from many institutions and individuals as well as the references and quotations from numerous reports and research results constitute the back bone of this book. Individual's acknowledgement apart, we thank all those whose work is cited in or ideas are used as inputs. We are especially thankful to all those publishers of research material who granted the copy right permissions.

First of all, the authors are pleased to thank the Vice-Chancellor of Savitribai Phule Pune University for providing an opportunity to Dr. Pant to be associated with the Department of Atmospheric and Space Sciences while he worked on this project. The most important support has been given by the Director General of Meteorology, IMD, New Delhi; Director, IITM, Pune and the Director, ARIES, Nainital. The authors are grateful to them for their encouragement to the author who joined this team from their respective organizations and made use of the accessible data from the archives of their institutions for analysis and studies incorporated in this book.

The authors would like to thank Dr. K. Rupakumar (WMO, Geneva), Dr. R. Krishnan, Dr. Ashwini Kulkarni, Dr. H. P. Borgaonkar and Dr. S. K. Patwardhan (IITM, Pune) for their advice, inputs and comments. Our special thanks are due to Prof. P. N. Sen, Dr. Rohini Bhawar (SP P une University) and Ashish Kumar (ARIES), for reading through the manuscript and providing critical comments and inputs. We received encouraging response and had informal discussions with many colleagues and friends. To mention a few, we acknowledge our thanks to Prof. Girijesh Pant (JNU), New Delhi, Dr. R. S. Tolia and Dr. Kusum Arunachalam (Doon Univ. Dehradun), Prof. D. S. Joag, Prof. K. C. Sinha Ray,

Prof. P. S. Salvekar and Dr. Anand Karipot (SP Univ. Pune). Our sincere thanks are due to Mr. Raman Solanki (ARIES) and Ms. Sonali Shete (SP Pune Univ.) for their help in creating some of the plots and figures. We wish to thank Mr. Anup Sah a renowned photographer from Nainital for providing some of his high resolution pictures taken during his tracking expeditions through the glaciers. It is heartening to note that one of his photographs of central Himalayan glacier has found place in the cover page. Thanks are also due to Mr. Nilendu Singh (WIHG, Dehradun) for his picture of Dokriani glacier, R. Chandra of Dainik Jagran (A Hindi Daily of Nainital) for a picture of disaster at Devli village and Mr. Manu Pubby of Indian Express for post devastation photograph of Kedarnath.

Authors would like to thank Mrs. Nisha Pallath for reading through some of the chapters and providing helpful inputs. Mrs. Preeti for reminding Dr. Narendra about the deadlines which helped in timely completion of the book and above all to Mrs. Gita Pant for providing many nature photographs from her collection for use as theme photographs of many chapters and also at few places in the text. The authors would like to record their gratitudes to their respective families whose support has been a source of inspiration throughout the course of this work. Last but not the least is our sincere thank and appreciation to the Springer Nature publication team for their coordination, patience and guidance throughout the manuscript preparation, editing and printing of the book.

Contents

Chapter 1
Climate and Climate Change: An Overview

1.1 Introduction

Climate is usually defined as the long-term aggregate weather conditions over a place which displays large variability in space and time. It is often said "Climate is what you expect and Weather is what you get." Climate of the Earth has distinct records of changes in the past and possesses potential to change in future on various temporal scales. The variability and change in the climate may occur due to natural reasons as well as those related to the anthropogenic (man-made) activities within the Earth system. Many changes in climate have been detected in the data relating

G.B. Pant et al., *Climate Change in the Himalayas*,
DOI 10.1007/978-3-319-61654-4_1

to the evolution of geomorphic features and life on the Earth within a time span extending back to the geological periods. The climate on these time frames is referred in a broader sense like; the colder/warmer periods or the wetter/dryer epochs, than the precise definition of climate of the recent past. The archeologists and historians have attributed many episodic changes in human civilizations and their settlements to major climate shifts like persistent droughts and floods, glacier surges, and the advances and retreats in desert margins across many parts of the globe. The climate induced abrupt changes in the human civilizations of the past might have occurred through the routes of environmental stress on basic needs of life: food, water, and shelter, as well as the consequences of natural disasters.

During the recent decades, the widespread warming of the Earth's surface, ocean waters and melting of large volume of ice on global scale, reported by authentic observations, have established that there are causes for these changes beyond natural processes. The present knowledge and understanding on climate science confirms that these changes have enough potential to create an imbalance in the Earth's climate system. Continuous monitoring of many climatologically significant parameters during the last about a century, as well as, a better understanding of mechanisms involved in the evolution of climate and changes there in, have made it possible to distinctly visualize the contemporary climates. All timescales ranging from seasonal to millennium and beyond can be used in defining the climate depending on the context. However, to define the present day climate in terms of surface meteorological parameters the climatologists generally consider a period of 30 years to represent the climate.

The climate change is thus recognized as a phenomenon that is being experienced by the people all over the world as a process with high potential to create perceptible changes in the Earth's ecosystem. There are indications that the variability and trends in global and regional climates have started appearing as major threats to traditional subsistence agriculture, water resource, clean air, and human safety, thus causing widespread material and social insecurity. This concern may assume significant dimensions particularly over marginal geographic zones over the Earth's surface, where higher sensitivity to changing climate is distinctly displayed. Small islands, coastal areas, arid regions, mountains, and glaciated regions of the world are under the category of high sensitivity zones for climate change. Within these zones there will always be an assembly of vulnerable sections of society likely to be affected the most, primarily due to enhanced stress on natural resources and inadequate socioeconomic protection.

India experiences a large variety of climates all across its latitudinal and longitudinal spreads, with warm-humid tropical climate in the east to an arid climate in the west and in a similar manner tropical monsoon climate in the south to temperate climate of middle latitudes in extreme north. The large-scale ocean-atmosphere coupled systems like the monsoons of south Asia are potentially sensitive to major climate change episodes, thus making India as one of the highly vulnerable regions due to its dependence on monsoons for practically all socioeconomic activities. It is, therefore, logical to assume that the direct impacts of climate change may adversely affect vital sectors such as biodiversity, forests, water resources, deserts, coastal regions, monsoons, mountains, and glaciers of the Indian subcontinent.

To deal with these adverse consequences of changing climate, a basic understanding on the subject is highly desirable for wide spectrum of educated society in the present day world. In view of this, an attempt is made in this book to cover the important basic scientific facts and other related issues pertinent to climate and its change. While dealing with issues relating to the social and environmental impacts of climate change, a balanced view is presented with supporting data and observations wherever possible. The subjects of vulnerability, impacts, mitigation, and adaptation are discussed as a regional case study for Himalaya and its subregion, the Central Himalaya (CH), keeping in mind the sensitivity of the region to changes in climate and a large number of concerned reader community.

The first few sections will deal with the story of the evolution of the science of climate change and its theoretical and observational foundation. A panoramic view of chronological developments in scientific and technological advancements in the field of theoretical basis of climate science as well as the observational methods used in the measurement of a large assembly of relevant parameters are also presented. The contemporary developments on these two important aspects have been complementary to each other and the growth in the recent past has been phenomenal.

As a follow-up on detailed discussion of climate and climate change, the second part of the book focuses on a highly significant region over the Asian land mass, the mighty Himalaya, primarily the Indian domain over the western sector, henceforth referred as the Western Himalaya (WH). This will cover an elaborate description of the climate of the Himalaya and its role in global climate, signals of regional climate change, and the probable impacts of global climate change on the overall mountain ecosystem. In the third part, a detailed description of climate change impacts will be described specifically for the CH region primarily represented by the state of Uttarakhand in the northern part of India.

As an introduction to the overall theme, following few paragraphs are devoted to present a concise view on the scientific basis and mechanism of climate change. It is well known that the natural changes in mean climate are primarily due to the internal dynamics of climate system as well as the changes in external forcings caused by the variability of energy input from the Sun. All planets of the solar system including the Earth and their satellites receive their energy from the Sun which is a radiant star located at the focus of orbital planes of these planets. The Sun emits its energy in the form of electromagnetic waves mostly in visible, ultraviolet, and infrared bands of the solar spectrum at an approximate surface temperature of 5778°K from a mean distance of 1496×10^5 km from the Earth.

The energy emitted by the Sun is enormous; however, a small fraction of it received by the Earth in the form of incoming solar radiation is the prime mover of all activities on the Earth including climate. This energy is defined by a term called the Solar Constant which is a measure of electromagnetic energy received from the Sun over a unit area on a plane perpendicular to the incident rays at a distance from the Sun equal to the mean distance between the Sun and the Earth. Though, the term Solar Constant is traditionally used to denote net energy flux with a fixed long-term average value, it displays minor variations due to changes within the Sun, such as the occurrence of the sunspots. Recent satellite measurements suggest a mean value of 1361 Wm^{-2} to the Solar Constant for all practical purposes including the study of climate change.

Similarly, the planet Earth, the only astronomical object known to accommodate and sustain life, emits energy corresponding to the temperatures acquired by all of its components according to the fundamental laws of radiation. This energy is much smaller in magnitude compared to the energy emitted by the Sun by virtue of the equilibrium temperature of the Earth [estimated to be 288°K (15 °C)] being less than one order of magnitude to that of the Sun. The radiant energy emanating from the Earth lies entirely within the range of infrared part of electromagnetic spectrum commonly termed as outgoing longwave radiation (OLR) and have a mean outward flux of 390 Wm^{-2}.

The energy received from the Sun as incoming shortwave radiation (visible) and that emitted by the Earth's surface as OLR (infrared), and the fluxes of sensible (physically measurable) and latent heat (involved in phase change processes without changing temperature) energies have a balanced budget within the Earth system. As a result, under normal conditions the warmth bestowed on the planet will maintain a constant value, as net loss of heat energy to outer space is adjusted in such a manner that the Earth's surface would tend to acquire a mean equilibrium temperature. In case the balance requirements of incoming and outgoing radiations are not fulfilled, the Earth's surface temperature would also change accordingly so as to eliminate the imbalance. For the basic understating of a common reader, the general picture of the electromagnetic spectrum is displayed by Fig. 1.1 given here.

Any change in the equilibrium thermal state of the Earth over a period of time resulting into change in its surface temperature and all forms of its manifestations on the Earth system are synonymous with climate change. Therefore, the commonly used index to quantify changes in climate of the Earth is the change in the magnitude of global mean surface temperature averaged over long period of time with respect to a standard base value. A persistent increase in the value of this index noticed during the last few decades is popularly known as the Global Warming (GW). It is thus a pertinent question for us to seek logical and scientific answers to observed changes

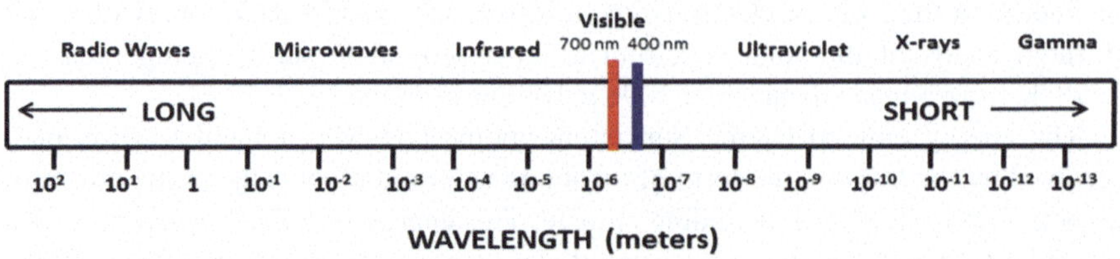

Fig. 1.1 Spectrum of solar radiation, colored portion representing the visible wavelengths that reaches up to the earth's surface

in the thermal state of the Earth we live in; and more puzzling for the scientists to find causes, understand mechanism, quantify them in terms of relative importance of different factors and predict the future course.

To understand and appreciate the operating mechanism of the climate system (discussed later), it is essential that the basic structure of the Earth's atmosphere up to about first 50 km (significant from climate point of view) is also briefly introduced in the beginning along with the basic concept of the radiation balance explained above. The lowest two layers of the atmosphere relevant for understanding of climate change are the Troposphere and the Stratosphere. Troposphere is the lowest portion of the earth's atmosphere with mean thickness of about 12 km (height is shorter near the poles and longer near the equator) in which the temperatures decreases upwards at an average rate of 6.5 °C/km^{-1}, known as the mean temperature lapse rate in the atmosphere.

The layer just above the Troposphere is known as the Stratosphere which starts from the region called the Tropopause, a thin isothermal (constant temperature) layer, separating the two. The stratospheric column occupies the region of the Earth's atmosphere approximately with a mean value between 14 and 48 km in which the temperature increases with altitude. The Stratosphere is characterized by the presence of ozone layers which absorb the ultraviolet component from the incoming solar radiation and protects the Earth's surface from these lethal radiations. The troposphere contains almost 75% of the total mass and about 99% of all the water vapor and aerosols of the atmosphere. Thermal structure of the atmosphere is given in the Fig. 1.2 below.

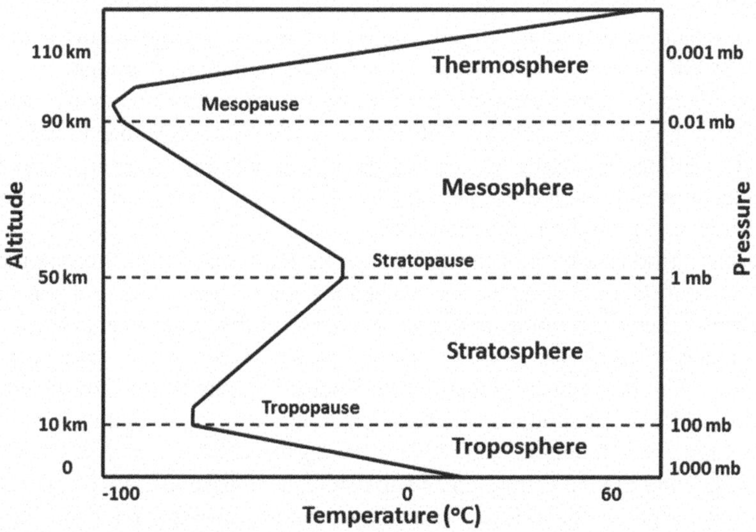

Fig. 1.2 Variation of temperature with height in the atmosphere

A very small fraction of the atmospheric GHGs and other pollutants from below may sometimes inject the basic input gases to the stratosphere. Minor contributions to stratospheric gases also come from the emissions of high flying aircraft, pollutants injected by volcanic eruptions, and the water vapor carried by deep convective cloud tops or transported through the atmospheric jet streams. Some of these gas molecules act as catalysts to the photochemical reactions involved in the formation as well as the destruction of stratospheric ozone. The abnormal increase in the concentrations of these anthropogenic gases is linked to the destruction of stratospheric O_3 which may increase the intensity of ultraviolet radiation (UV) reaching the surface of the Earth. An example of this is seen in the form of an event of depleting O_3 over the Antarctic region, popularly known as the Ozone Hole. This phenomenon is related to the creation of O_3 deficiency over the region due to the accumulation of O_3 depleting substances whirled in by semipermanent polar stratospheric circulation systems.

During the recent geological phase of the Earth, most of the observed increase in temperature of the Earth's surface has been occurring since the mid-twentieth century. This unusual happening is attributed to monotonous increase in the concentration of anthropogenic GHGs which absorb the outgoing thermal radiation from the Earth by virtue of their molecules behaving differently to this radiant energy than the molecules of other major constituent gases of the atmosphere. Increasing trend in the concentration of these gases, particularly the carbon dioxide, methane, and nitrous oxide is directly related to the enhanced burning of fossil fuel and the land use and land cover changes which have occurred during the last few decades as a result of human activities (IPCC-WG-1 2013).

It is interesting to note that the presence of normal amounts of GHGs in the atmosphere is highly significant for maintaining the warmth on the Earth without which it would have been much cooler with a mean surface temperature of approximately 255°K (−18 °C). In a holistic sense, the perturbations to climate system by anthropogenic influences are often referred by the scientific community as alterations in the biogeochemical cycle of the Earth. The biogeochemical cycle in the context of climate system is defined as the cyclic pathway by which a chemical substance moves through biotic (biosphere) and abiotic (atmosphere, hydrosphere, cryosphere, and lithosphere) components.

Having introduced in brief the climate of the Earth and its functioning through the process of the radiation balance, we embark on the presentation of some basic introductory material relating to ground realities over the target area, the Himalaya. The Himalaya, a great mountain barrier spread across latitudes in the heart of the Asian continent poses many scientific challenges to climatologists for understanding and predicting the future course of changing climate over this region. The Himalaya also has a unique distinction of being the vast land area with relatively smaller quantity of GHG emission and an equally extensive forest cover to provide major sink for carbon dioxide, the main player in the global warming game. Except for the coastal regions all other climate-sensitive zones stated above are a part of the Himalayan belt. It has high level of biodiversity with many hot spots and endangered biological species and an extensive and diverse forest cover including grasslands and prairies. The Himalaya is a permanent reservoir and perennial source of freshwater to major rivers and also feeds to enumerable glaciers of the country. Finally, the Himalaya also has some desert areas in the form of cold-dry plateau regions of north Kashmir and Himachal Pradesh. The mountains of the Himalaya with extensive permanent snowcap on their top have complex topographic features with highly variable geography, geology, biology, and the habitats.

The summer monsoons of South Asia and the weather systems of the middle latitudes provide rains, snow, hail, lightening, thunder, and devastating furry of extremes of weather to the region which sustain and endanger the fragile mountain ecosystem. The Himalaya and the Tibetan plateau to the north of it play a dominating role in influencing the monsoon systems of south Asia which is a subject of great scientific significance to the climatologists. India is fortunate to have this vast and unique open space for observation, experimentation, simulation, and understanding of the behavior of changing climate and its environmental impacts. Keeping in mind the role of the Himalaya in global and regional climate, a major part of this book deals with its climate, the signals of change, and their impacts on the sustainability of the total mountain system dynamics under changed climate scenario. For the purpose of detailed climate change analysis, an attempt has been made to analyze the long period data for the WH, consisting of regions in Jammu Kashmir, Himachal Pradesh, and Uttarakhand states in north and north-west India.

The last and the most significant attempt in this book is to present the climate change signals, their detection, and impact analysis for a section of vulnerable geographical segment identified as the CH. The CH, major portion of which consists of the state of Uttarakhand located in north-central India, provides a unique example of the most sensitive region for climate change impact study. The region experiences monsoon climate with good amount of rains over large area in the plains and the southern slopes of high rise mountain ranges during the south-west monsoon season. It also encounters the snow and rain bearing weather systems of middle latitudes entering in sequence from the west over the Indian latitudes during the winter months, known as the western disturbances (WDs). The post- and pre-monsoon seasons in the region are equally eventful with sporadic and occasional but often vigorous weather events. Data on vital climate-related parameters over the CH are very sparse and available for much shorter time periods over most of the region except for few selected stations which is presently inadequate for a comprehensive climate change study. The satellite measurements of few basic parameters generated through globally interpolated and assimilated data sets are available for the last few decades and are being extensively used by the climatologists to study the broad climate change features.

The region thus demonstrates a dynamic interplay of all forms of weather and climates interacting with a highly fragile mountain ecosystem to support and sustain its diversity and periodically experiencing their disastrous consequences. It is therefore an obvious choice for the study and description of climate change impacts and related socioeconomic and policy matters as a typical case study in the broader context of future changes in the Himalayan climate.

1.2 Science of Climate and Climate Change

Strictly in a scientific sense the climate is a description of mean equilibrium state of the Earth system which displays significant variations in space and time. As explained in the previous section these variations primarily arise due to the differences in the amount of incident solar energy inputs across latitudes as a function of time and space and local geography, such as surface morphology, land-ocean contrast, vegetation cover, and the altitude. Historically, the climatologists have used long series of meteorological data in association with spatial distribution of vegetation regimes and generated sets of classifications to delineate distinct climatic zones over the Earth's surface.

The delineation of climate zones has been based on the concept that vegetation is the primary representation of climate, thus recognizing the role of climate in establishing major biospheric activities. Therefore, the boundaries of climate zones are demarcated with vegetation distribution as basic information along with weather factors such as the average annual and monthly temperatures and precipitation and their seasonality. Scientific methods of classification developed in the early years of twentieth century were further modified by incorporating the role of additional parameters like evapotranspiration, relative humidity and extremes of temperature, and precipitation. The climate science has thus emerged as an amalgamation of physical and life sciences studied in an interdisciplinary environment as a major discipline of the Earth system science.

The pioneering contributions of a large galaxy of scientists from diverse disciplines have laid the foundation of climate science over a span of many centuries. It is therefore appropriate to mention a few among the galaxy of classical biologists, astronomers, geographers, and the climatologists to make sure that their contributions are not lost in oblivion. To name a few among the pioneers are Aryabhatta (Indian, 476–550 AD), Varahamihira (Indian, 505–587 AD), Brahmagupta (Indian, 598–670 AD), Alexander Von Humboldt (Prussian, 1769–1859 AD), Vladimir Peter Koppen (Russian/German, 1846–1940 AD), Milutin Milankovitch (Serbian, 1879–1958 AD), Rudolf Geiger (German, 1894–1981 AD), Glenn Thomas Trewartha (American, 1896–1984 AD), Charles Warren Thornthwaite (American, 1899–1963 AD), Helmut Erich Landsberg (German, 1906–1985 AD), Hubert Horace Lamb (British, 1913–1997 AD), and Mikhail Budyko (Russian, 1920–2001 AD).

With improved understanding and availability of long series of data on many parameters, the climate at present is in the realm of most sophisticated science disciplines of social relevance. As a result, it is now possible to precisely explain and model the physical, chemical, and biological processes within the planet Earth, which constitute its climate and their mutual interactions and also to study their time-dependent behavior. The information thus generated will have the vital clues about likely future states of the Earth's environment. This information will prove to be of vital importance in guiding the planning process for sustainable developmental activities of human societies as well as a helpful tool for the minimization of losses due to climate-related disasters.

A holistic picture of global climate emerges out of evolutionary and interactive processes within a complex system (climate system) consisting of five major components, namely, Atmosphere (thin gaseous envelope around the earth), Hydrosphere (all water forms), Lithosphere (surface skin of solid earth), Cryosphere (all ice, snow, and frozen water mass), and the Biosphere (all living beings on the Earth's ecosystem). This system is perennially driven by an uninterrupted source of energy in the form of incident radiation from the Sun (radiative forcing). Therefore, the climate with this comprehensive definition is understood as a grand manifestation of state of the Earth as a whole, averaged over periods of time long enough to cover variations in statistical properties of relevant quantities representing these states.

During a period of about half a century of postindustrial revolution and enhanced developmental activities all over the globe, the climate system has witnessed significant modulations mostly resulting from direct human interference. As a result, a new component relating to human interference, represented by a recently coined term "Anthroposphere" has been added to the climate system. This incorporates the entire spectrum of human activity and consequent perturbations in natural biogeochemical cycle. The period of last about 10,000 years which has witnessed different degrees of human interference in the climate system is designated as an era in the geological history of the Earth, appropriately termed as "Anthropocene."

The geologists have identified distinct glacial interglacial phases in the recent climate history of the Earth considering the prevailing temperature regimes as one of the important criteria for their definition. During the most recent geological period called the Quaternary since about a million years ago, there are many instances of prolonged climate extremes in terms of large temperature and precipitation anomalies (significant departure from long period mean condition) over the Earth's surface. The climate is therefore characterized by its unique variability in time and space with differing magnitudes and signs and interpreted accordingly by the practitioners of variety of scientific disciplines to suit their requirements. The concerned groups include the social scientists and policy makers in addition to climatologists and a plethora of multidisciplinary earth system scientists.

The climate system thus experiences a state of fragile equilibrium which has been subjected to natural changes due to external factors. These changes occur through variability in the solar energy input arising out of periodic changes in astronomical position of the Earth in the solar system, feeble changes in solar energy due to sunspots, solar flares, cosmic rays, and the magnetic fields. In addition, the changes may also take place by the interruption of incoming solar energy due to gaseous particles ejected into the atmosphere from the volcanic eruptions.

The astronomical influences commonly known as the Milankovitch theory named after the Serbian geophysicist Milutin Milankovitch, who worked on the idea as a Prisoner of War during the World War-I, are the primary causes of cyclic changes in climate over a time period of the order of thousands of years. His theory describes the effect of changes in the Earth's movement in its orbit around the Sun and their collective effect upon its climate (orbital forcing) due to small but significant changes in the amount of solar radiation reaching the top of the atmosphere over a long period of time.

It is very fundamental to understand the functioning of the orbital forcing in terms of the natural changes in the Earth's orbit around the Sun. The orbit of the Earth which is elliptical in shape varies in time between near circular to mild elliptical with an approximate cycle of 98,000 years and is known as the eccentricity of the orbit (orbital shape). This influence is due to the changes in the gravitational interactions of the Earth with the planets Jupiter and Saturn. The axis of the Earth is also tilted with respect to its orbital plane which oscillates between 22.1 and 24.1° in a time period of approximately 42,000 years and is known as the obliquity (axial tilt). In addition, the axis of rotation of the Earth has a spin motion with respect to the fixed star with a rough periodicity of 21,000 years and is known as the precession (axial spin). This spin arises due to changes in the gravitational influences of the Sun and the Moon on the oblate spheroid shape of the Earth. A schematic representation of the orbital effect is presented in Fig. 1.3.

The combined effects of these three contribute to the changes of the scales of glacial interglacial phases (many thousands of years) in the climate of the Earth. The data from cores drilled by the geologists from deep sea sedimentary deposits have provided direct evidences of glacial and interglacial phases which the climate of the Earth have witnessed in the recent geological past. There are unambiguous conclusions that these changes are substantially influenced by the orbital forcing and are the authentic imprints of the past global changes (Hays et al. 1976; Muller and McDonald 1997).

As stated above, it is important to note that the climate evolves out of a set of numerous nonlinear processes and interactions among various components that determine the fluxes of mass, energy, and momentum near the Earth's surface and also the interlinking feedback mechanisms. The nonlinearity in a complex multi-

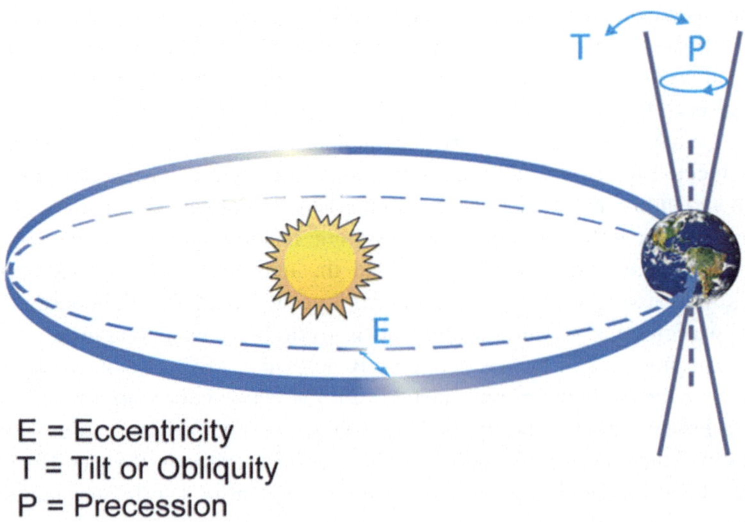

Fig. 1.3 A schematic representation of the Milankovitch hypothesis (source: FAQ 6.1, Figure1 of IPCC-2007, p. 449)

variate system like climate can be defined as a relationship between the variables which cannot be explained as a linear combination of its inputs to determine the cause–effect relationships. Though, it is possible to model and simulate mean climate states and their properties reasonably well, the uncertainties introduced by observational errors and nonlinear and chaotic behavior of some of the governing processes, the climate has inherent limitations on its predictability.

Therefore, the climate predictions such as to foretell the distribution of global temperature or the rainfall patterns a decade or 50 years from now, are generally confined to probabilistic estimates with designated uncertainties. These projections of future climates are generally spelled out in terms of the mean and range of variability as well as the trend of change in the values of standard meteorological parameters particularly, the temperature and precipitation. In addition to the trend of change in mean climate, generally calculated as a best fit linear estimate, there are large variations on seasonal, interannual, and decadal scales which need to be accounted for. Major component of these variations emerge out of unique occurrences of weather patterns associated with the large-scale circulation systems across the globe. The prominent among these are the El-Nino (unusual warming of ocean surface waters along the east-central equatorial Pacific), monsoons, and also a combination of numerous synoptic and mesoscale weather systems and their random spatial and temporal patterns.

The most recent examples of distinct epochs defined by detectably different climates are the Last Glacial Maximum with its peak around 18,000 years before present (ybp) followed by the Medieval Warm Period witnessed around the period 900–1300 AD. The most recent episode observed around the period 1400–1800 AD is known as the Little Ice Age. The changes in climate which have occurred in the past can be detected or monitored by measuring relevant physical quantities or the proxy indicators of changes in temperature and precipitation. Over the South Asian monsoon region, there are historical evidences of these periods in the form of social impacts of droughts over India though they need to be corroborated with physical paleoclimatic evidences (Mooley and Pant 1981).

Change in climate due to natural forcings can be explained in terms of cause–effect relationships and can be measured and calculated in quantitative terms. However, the changes attributable to the alterations of biogeochemical cycle by anthropogenic activities may carry inherent uncertainty due to the constraints imposed by the unpredictable behavior of those activities. The anthropogenic activities are usually governed by socioeconomic, geopolitical, and policy considerations. The increase in surface temperature of the Earth due to the anthropogenic influence on climate system through radiative interactions (alterations in radiation budget of the earth-atmosphere system) involving atmospheric GHGs may concurrently occur along with the changes due to the natural causes. Therefore, it is a great scientific challenge to separately account for these components of climate change and distinguish them in a complex embedded system such as the climate.

It is now well established that the anthropogenic interference in climate system is primarily through the emission of the GHGs. The major GHGs in the atmosphere are carbon dioxide (CO_2), methane (CH_4), nitrous oxide (N_2O), tropospheric ozone (O_3), water vapor (H_2O), and the chlorofluorocarbons (CFCs). The other oxides of nitrogen, nitrogen dioxide (NO_2) and nitric oxide (NO), together known as nitrogen oxides (NO_x) are two other species of atmospheric gases which play a crucial role in the formation of O_3 and secondary aerosols (minute suspended particles) in the atmosphere. The GHGs in the atmosphere absorb the infrared thermal radiation emitted by the Earth and increase the temperature of the Earth's surface and lower layers of the atmosphere similar to the warming in a greenhouse chamber in an agricultural farm.

Most of these gases once released into the dynamic atmosphere, travel across and mix with the atmospheric air and become part of the atmospheric circulation system with varying degree of residence times and GW potentials (GWP). The GWP is an index of warming on global scale from different sources responsible to induce the radiative forcing. The GWP is defined as an index based on radiative properties of GHGs, measuring the radiative forcing following a pulse emission of a unit mass of a given GHG in the present day atmosphere integrated over a time frame (e.g., the Kyoto Protocol is based on a time frame of 100 years), relative to that of CO_2. Most important among the GHGs, the CO_2 thoroughly mixes with the atmospheric gases, and has a relatively longer residence time in the atmosphere. Scientifically monitored and calibrated measurements from many parts of the globe suggest that the concentration of CO_2 in the global mean atmosphere have increased from preindustrial value of about 280 parts per million (ppm) to its current value of about 410 ppm as of today and it is demonstrated to follow an exponential profile. The global atmospheric concentrations of other two important GHGs, the CH_4 and N_2O, have also increased in similar fashion during the same period of time and under similar conditions.

Proxy data on past climate suggest that the present concentrations of the atmospheric GHGs, CO_2, CH_4, and N_2O, is unique and unprecedented since it exceeds the range of concentrations recorded in ice cores during the past 800,000 years (IPCC-WG-1 2013). Recent research have confirmed that the optical properties of aerosols and black carbon (carbon shoot) suspended in the atmosphere also possess the potential of contributing to anthropogenic changes in the climate. These changes take place through radiative interactions as well as through their role in inducing changes in the microphysical properties of clouds. Observational evidences and the Global climate model studies suggest that there is a net negative contribution to the radiative forcing (cooling) from most aerosols and a positive contribution from the absorption of the solar radiation (warming) by black carbon in the atmosphere.

The fundamental concept on selective absorption of radiant energy by different materials subsequently used to explain the greenhouse effect in the atmosphere is more than almost four century old. Edme Moriotte (French, 1620–1684 AD) in the year 1681 AD noted that the Sun's rays with light and heat pass through glass and other transparent material but the heat from other sources does behave differently. Horace Benedict de Saussure (Swiss, 1740–1799 AD) conducted an experiment in 1760 AD using an instrument named Heliothermometer (panes of glass covering a thermometer in a dark box) to demonstrate an analogy to the then existing concept similar to the greenhouse effect. Taking a clue from Saussure's experiment, famous mathematician Joseph Fourier (French, 1768–1830 AD) first introduced the concept in the year 1824 AD that the Earth is not heated so much by direct solar radiation as by longwave radiation emitted by the ground.

Advancing this idea further, the well-known Swedish chemist Svante August Arrhenius (1859–1927 AD, Nobel Prize-Chemistry-1903) was the first to make an attempt around the year 1895 AD to quantify the effect of changes in atmospheric CO_2 concentrations up on the temperatures on the Earth's surface. In a paper published in the Proceedings of the Swedish Academy of Sciences (Arrhenius 1898), he suggested that 40% increase or decrease in the abundance of atmospheric CO_2 might trigger the interglacial and glacial periods. His calculations on global warming using limited data and simple assumptions are comparable to the estimates obtained on the basis of the recent data. His explanation about glaciations was later on modified incorporating influences due to astronomical positioning of the Earth in its orbit being partially responsible for modifying the solar energy input which is argued to be responsible for triggering of the glacial interglacial episodes.

These initial ideas were subsequently pursued by many physicists who conducted experiments aimed at demonstration of the Earth's radiation balance and also to measure and quantify the heating effect. Claude Pouillet (French, 1790–1868 AD) studied the heat produced by the Earth and developed an instrument in the year 1838 AD called the Pyrheliometer to measure the solar radiation and determined the quantitative value of the solar constant. John Tyndall (Irish, 1820–1893 AD) in the year 1859 AD conducted laboratory experiments to demonstrate the absorption of thermal radiation by complex molecules as opposed to the primary bimolecular atmospheric constituents, oxygen and nitrogen.

After these pioneering experiments for almost about a century, many discoveries emerged as a result of the studies relating to the entire spectrum of electromagnetic radiation. These research efforts spanning over to almost a century involved a large number of astronomical and spectroscopic measurements and theoretical calculations on the physical characteristics of electromagnetic radiation. These scientific investigations and fundamental discoveries have established a hierarchy of fundamental laws of radiation resulting into many Nobel Prizes in physics being awarded to many of these scientists, which encouraged further research and experiments in the field. Most of these discoveries have direct scientific relevance to functioning of the Earth system and created the foundation of subsequent research and observations highly relevant for climate science.

Readers may find it interesting to recall the winners of prestigious Nobel Prizes in physics, chemistry and engineering whose work and interest relating to radiative transfer and other topics have great relevance to the science of climate change. Important among these are Swante August Arrhenius (1903), Lord JWS Rayleigh (1904), Sir William Ramsay (1904), Albert Michelson (1907), Guglielmo Marconi and Karl Ferdinand Braun (1909), J. D. Vander Waal (1910), Wilhelm Wien (1911), Max Plank (1918), C. T. R. Wilson (1927), C. V. Raman (1930), Sir Edward Appleton (1947), William F. Libby (1960), Malvin Calvin (1961), S. Chandrasekhar (1983), Paul J. Crutzen, Mario Molina, and F. S. Rowland (1995).

Though, the first half of the twentieth century has been a period of modern scientific revolution particularly in Europe in the field of electromagnetic radiation; however, not much of attention could be paid specifically to the radiative processes in the Earth's atmosphere. Till about last half a century, it was believed that CO_2 and H_2O are the only GHGs in the atmosphere taking part in the radiation balance of the

Earth's climate system. Recent enthusiasm in climate change research has motivated the scientific community to look for new atmospheric gases and other secondary products of atmospheric chemical processes with an eye of a detective to ascertain their GWP and include them in the family of GHGs.

Numerical modeling of the atmospheric effects of changing CO_2 through radiative processes started with the first experiment by Guy Stewart Callendar (British, 1898–1964 AD) in 1938 (Callendar 1938; Hawkins and Jones 2013). He solved a set of equations linking GHGs to climate and demonstrated that doubling the atmospheric concentrations of GHGs in the equations resulted into an increase in the mean global temperature of the earth's surface by 2 °C with enhanced warming rates near the poles. The developments of last about half a century have made it possible to quantitatively simulate and model climate processes at land, ocean, and atmosphere including their interfaces to develop dynamic mathematical models to predict future climates with a set of initial conditions.

These models are developed using the fundamental laws of conservation of mass, momentum, and energy in a rotating Earth system with time-dependent set of basic equations and also take into account different external forcings. These developments have opened the flood gates of multidisciplinary theoretical and observational research in climate science in which the lion's share was devoted to the physics, chemistry, and modeling of GHGs, more specifically, CO_2, other GHGs, aerosols, cloud microphysics and the boundary layer processes. To examine the impact of doubling of CO_2 in the equilibrium climate and to study sensitivity of models to changing CO_2 levels using Global Atmospheric General Circulation Models (AGCMs) became a priority research topic in the decade of 1980s (Schlesinger and Mitchell 1987; Meehl and Washington 1993). Subsequent developments in the modeling of climate incorporating coupling of land, ocean, biosphere, cryosphere, aerosols, and clouds with atmosphere are formulated to analyze the model sensitivity to GHGs as a basis for future climate change scenario generation.

Research and development in the field of climate change specifically the Earth system modeling and observations have witnessed phenomenal advancements with appreciable success during the last quarter of a century. It will not be possible to include in this book the details about different models and their results including a rapid advancement in the development of a hierarchy of climate models and their success story. The account in this book is therefore constrained to have a restricted view on the subject primarily due to the choice of target readers and the priorities to other aspects. However, a brief introduction at a later stage is attempted to provide an overview of the admirable developments of the last few decades.

It is also necessary at this stage to introduce in simple terms the role of the GHGs other than CO_2, as listed earlier, in the process of GW. The water vapor is the most abundant GHG in the atmosphere though it is not possible to directly incorporate its change in climate process and to measure quantitatively the trends of change in net water content of the atmosphere. Water in the atmosphere is inducted and circulated through a combination of multiple agencies such as vertical ascent, horizontal advection, evaporation from surface sources, and transpiration from flora and fauna at the biosphere level as well as dynamic weather processes within the atmosphere.

The water in climate system in its totality is defined in the form of a recycle loop termed as the hydrological cycle, encompassing all water sources, sinks, and forms. Water in the atmosphere exists in three phases: solid, liquid, and gas in which an active phase change mechanisms may operate any time anywhere in the first few kilometers of the atmosphere (troposphere) and the vertical ascend results in all sizes and shapes of the hydrometeors (water droplets, hailstones, and ice crystals). The latent heat content of water in the atmosphere provides a critical driving force for atmospheric circulations on scales ranging from individual thunderstorms to the global scale circulation systems.

The total water carrying capacity of the atmosphere is a function of temperature; therefore, it has a cap on its upper limit since a fully saturated atmosphere with a given temperature profile will not accept more water to be assimilated in it under normal conditions. Water content of the atmosphere is highly variable in time and space; though, for an average thermal distribution its integrated value over the globe may remain constant. Changes in concentration of water in the atmosphere can thus be considered as a resultant product of climate feedback due to water vapor and clouds related to thermal state of the atmosphere rather than a direct input from anthropogenic sources. Climate models generally take into account the presence of water as a conserved quantity like; mass, energy, and momentum while recognizing the corresponding increase in the total water carrying capacity of the atmosphere as a consequence of changes in mean surface temperature.

Methane is the second most prevalent GHG in the atmosphere after CO_2. It is emitted by natural sources such as wetlands, natural gases, waste material, and organic decomposition over land and the water. However, the larger part of methane is generated globally from anthropogenic activities such as industry, agriculture (manure, soil, rice fields, livestock digestive processes, etc.), and waste from homes, industries, and business. Certain natural processes in soil and near the ground along with chemical reactions in the atmosphere also help in the removal of CH_4 from the atmosphere. The residence time of CH_4 in the atmosphere is much shorter and its quantity is much less than that of CO_2 though it has a much higher efficiency in trapping the heat of the Earth's surface from escaping out; therefore, its contribution is considered quite significant as a GHG.

Nitrous oxide is naturally present in the atmosphere in very small quantity as a component of the nitrogen cycle of the atmosphere originating from a variety of natural sources over the land and oceans. Many anthropogenic activities such as agriculture, fossil fuel combustion, wastewater treatment and certain industrial processes are responsible for the increase in the amount of N_2O in the atmosphere. As a part of nitrogen cycle, they are also being destroyed through chemical reactions with the residual amount staying for a long time and possessing a much higher GWP than the equivalent amount of CO_2. Even then the net GW effect is small since the relative concentrations of N_2O are much smaller compared to the CO_2 and CH_4.

The CFCs, an abbreviation for the organic compounds chlorofluorocarbon, commonly known with the brand name "Freon" are widely used as refrigerants, solvents, propellants, and sprays. The CFCs contribute to the depletion of ozone layer in the upper atmosphere. In addition to being an active ozone reducer, these anthropogenic compounds are also the gases with much higher potential to enhance the greenhouse effect than CO_2; though, the quantity involved is relatively small. Though, the net amount of CFCs in the atmosphere was very small compared to other GHGs at the time of its detection as a source of GW; in the meantime, it was noticed that the rate of increase was sharp due to its large-scale induction in variety of commercial activities.

These were considered with alarm primarily because of their being totally of anthropogenic origin. In view of their commercial applications, they carried a possibility of higher potential with accelerated growth and likely to cause irreparable damage to the environment in the near future. Fortunately, the manufacturing of these compounds have been phased out under the Montreal Protocol on O_3 depleting substances. This has encouraged the industries to look for other environment-friendly products as substitutes with good success thus, totally eliminating the CFCs at their source.

In addition, there are some short-lived GHGs such as O_3 and stratospheric water vapor injected from volcanic eruptions and high altitude aircraft and oxidation of CH_4 which contribute to anthropogenic radiative forcing. The O_3 is not emitted directly to the atmosphere but formed by the photochemical reactions involving precursor compounds, such as the NO_x of natural or anthropogenic origin. The tropospheric O_3 which is produced as a result of photochemical reaction involving pollutants and active radicals is very small and short lived compared to the stratospheric O_3, where the reactions persist within a photochemical chain. In the stratosphere, the role of O_3 is vital in absorbing the ultraviolet radiation though it has almost negligible role as an agent for radiative forcing. As mentioned earlier in connection with the tropospheric water, the water in the stratosphere (injected through volcanic eruptions, jet aircraft, and deep convective cloud tops) and its change are also considered as feedbacks rather than a forcing in the climate system.

Though, we are confining our discussions to the atmospheric constituents of the climate system, it is appropriate at this stage to have a brief mention of the role of the Ocean for completeness and highlight its significance in carbon, nitrogen, and water cycles. The oceans constitute the basic source and sink of water mass in hydrological cycle thus occupies an important position as an active partner in the climate change process. The Oceans are directly linked to radiative forcing through exerting substantial temperature-dependent control over release and sequestration of major GHGs like CO_2 and CH_4, in addition to being the major source of water and a dominating component of the climate system. The role of the Oceans in handling the CO_2 in effect is somewhat similar to the role of forests and vegetation in the sequestration of CO_2 over the land areas. Oceans are estimated to absorb more than 90% of extra heat produced through the anthropogenic activities. In fact, in an overall climate system process a very small amount of heat is available to the atmosphere for global warming.

Climate models usually couple the Oceans with atmosphere to generate reliable projections of climate parameters on longer time frame. The anomalies of sea surface temperatures, an important parameter used in Ocean-Atmosphere coupling evolving over a timescale of few years designated by the El-Nino phenomenon, are important climate parameters of the tropical oceans. These oceanic influences coupled with the atmospheric circulation systems of corresponding timescales are considered to have significant influence on monsoon climate over India which demonstrates the role of the Ocean-Atmosphere coupling on interannual scale variability. This influence is commonly described in the form of a combination of atmospheric circulation (Southern Oscillation) and the large-scale sea surface anomaly related to the El-Nino. Over the Indian and South Pacific region, it is denoted in a combined acronym called ENSO. On much longer timescales, the large-scale melting of polar ice sheets due to global warming is linked to the mean circulation systems of deep Ocean waters and altering the Ocean conveyor belts which is a mechanism to slowly mix the cold waters of the Polar region with the warm waters of tropical Oceans to maintain the long-term thermal balance of the Oceans.

1.3 Global Warming Trends

Comprehensive data on surface temperature of the Earth from all over the globe incorporating all possible sources at the command of climatologists covering a time period of more than a century is available for examining its secular trends. The data indicate a continuous increasing trend in mean temperature with an enhanced increase during the last few decades. It is seen that the maximum temperature in many parts of the globe shows an enhanced trend as compared to the minimum and mean temperatures. The global mean surface temperature data for more than a century suggest that each of the past four decades has been successively warmer at the Earth's surface than any of the previous decades in the instrumental records.

It is also reported that the decade of 2000s has been the warmest. The recent data points out that the year 2015 has been the warmest year with highest annual mean temperatures so far recorded. The trends of the summer months of the current year suggest that most likely the year 2016 may score higher and the trend may continue over the coming years. Observed data on global mean Earth's surface temperature combined for land and ocean together since the year 1850 AD compiled by the Intergovernmental Panel on Climate Change (IPCC) shows a monotonous increase during the period of last about 50 years as given by Fig. 1.4.

The linear trend calculated by using globally averaged combined land and the Ocean temperature data show a warming of 0.85 (0.65–1.06)°C, over the period 1880–2012 (IPCC-WG-1 2013) which roughly works out to be 0.65 °C for 100 years. These values for the 100 years period starting at 1901 are reported to be much higher to the tune of 0.74 °C (IPCC 2007; Du et al. 2004). These estimates may slightly vary in independently produced data sets and varying time periods used to calculate the global trends and also the methods used in calculating the trend value. They also display regional variability and temporal fluctuations with notable findings that the land areas warm more than the oceans and the higher latitudes experience more warming compared to the tropics. While examining the impact of GW, it should be remembered that the global mean values are arrived at after aver-

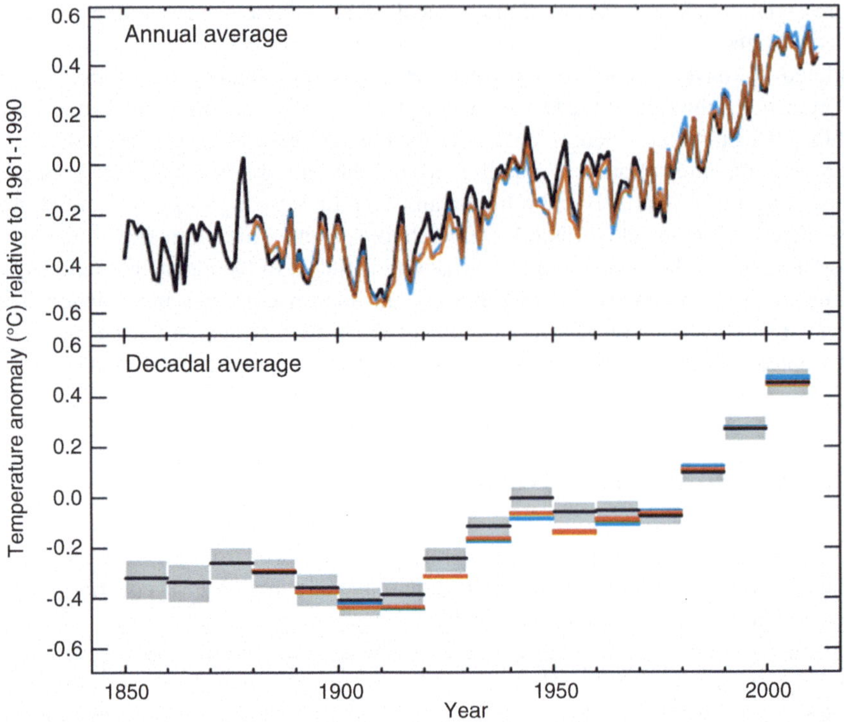

Fig. 1.4 Observed global mean combined land and ocean surface temperature anomalies, from 1850 – 2012 from three data sets. *Top panel*: annual mean values. *Bottom panel*: decadal mean values including the estimate of uncertainty for one data set (*black*). Anomalies are relative to the mean of 1961–1990 (IPCC-WG-1 2013, p. 6)

aging data with a large spatial and temporal variability. Therefore, it is possible that some parts of the globe may experience much higher anomalies than the values displayed in the global mean which may require special attention and concern.

These values for GW are supported by model simulated projections suggesting equilibrium climate sensitivity in terms of surface air temperature for the next 100 years in the range of 2.0–4.5 °C under different predetermined emission scenarios and model physics and configurations. The IPCC-WG-1 (2013) refers to the equilibrium/steady-state climate sensitivity (units: °C), as change in the annual global mean surface temperature following a doubling of the atmospheric equivalent CO_2. These estimates may slightly vary as new data is used to update the series and models generating future scenarios become more refined as the knowledge improves and remedial measures are adopted to control the emissions of greenhouse gases. Modeling studies suggest that once the global warming is triggered by the GHG increase, it will take many years to revert back to the original condition even if the emissions are controlled (IPCC-WG-1 2013), as the equilibrium state is reached through a combination of varying response times related to different components of the climate system.

1.4 International Concerns and Initiatives on Climate Change

The emerging scientific evidences and global dimensions of climate change and its likely impacts on all sectors of socioeconomic development, sustainable use of natural resources, and human safety has become a matter of global concern. A large number of governmental and non-governmental organizations (NGOs) all over the world have initiated serious dialogues and actions regarding remedial measures and strategy for adaptation to the change. During the last three decades, landmark agreements were reached and protocols were signed by many countries at multinational climate forums and differences have often surfaced on many contentious issues. These differences are mostly on the logistic and policy matters rather than the scientific facts, data, and information or the matters relating to regional perception on climate change.

Though, the central theme of many International activities is focused on wider arena of environmental issues, the climate has always been receiving utmost attention. The important milestones of success of these international efforts are the Montreal Protocol on Atmospheric Ozone (Canada, 1987) and the Kyoto Protocol on GHG emission controls (Japan, 1997) to mention a few. Public attention is also being drawn towards this important issue through galaxy of mega events, such as the UN Conference on Environment and Development held in Rio de Janeiro (Brazil, 1992) followed by a chain of recent international conferences, conventions, and official level summits. To address the issue in a scientific manner and to advise the participating member countries, the United Nations (UN) has established a permanent mechanism under the banner of the United Nations Framework Convention on Climate Change (UNFCCC), an international environmental treaty. The UNFCCC was proposed at the Rio conference in 1992 and came into effect on 21 March, 1994 with 196 member countries of the UN as parties clearly indicating an overwhelming support to the cause. An important forum under UNFCCC is the group called the Conference of the Parties (COP) which is the supreme decision-making body.

A key task for the COP is to review the national communications on emission inventories submitted by the member parties which is mandatory to all. The COP usually meets once a year to establish the rules and regulations and monitor the implementations. The most recent in the series, the COP-21, was held in Paris starting 30th November 2015. Though the outcomes of this event are not yet very clear, it is assumed that there are no emission reduction targets likely to be fixed. The emission reduction targets are being placed in the climate action plans. The strategy is revolving around a new term phrased as, Intended Nationally Determined Contributions (INDC), that every country is framing and putting forward and they are open to the decisions of the individual nations.

It is important to note that the concerns of a large number of member countries based on the implications of climate change have been recognized and deliberated upon by the UN and its various agencies much before the formalization of the

UNFCCC treaty in 1994. The IPCC, another very significant initiative by the world body, was established a few years earlier in 1988 under the auspices of the World Meteorological Organization (WMO) and the United Nations Environment Programme (UNEP). The scientific studies and analysis so far carried out by the IPCC along with its recommendations are most pertinent to the subject matter dealt here; therefore, some basic introduction to IPCC is given in the following paragraphs. The main objectives of the IPCC are to carry out the following tasks for which three different working groups comprising eminent global experts are established.

i. Compilation and assessment of available scientific information on climate change
ii. Assessment of environmental and socioeconomic impacts of climate change
iii. Formulation of response strategy including adaptation and mitigation measures

Since their inception, these working groups are engaged in respective tasks by involving multidisciplinary scientific community in their deliberations. As resultant products of these efforts, they bring out comprehensive reports on the above-stated three aspects once in every 5 year. The most recent in the series is the IPCC-5th assessment report released in the year 2013. The status and recognition of the IPCC reports and recommendations have received further impetus since the award of Nobel Peace Prize to IPCC which was shared with Mr. Al Gore of the United States, in the year 2007.

The assessment made by working group-1 on the Science of Climate Change provides the fundamental inputs and are the key to further analysis, interpretation, and actions at national and international levels. The other two groups extensively deal with the issues of emission control, mitigation measures, adaptation strategy, and remedial measures and to make recommendations on socioeconomic aspects and policy matters. One of the major conclusions of IPCC on the Science of Climate Change explicitly mentions the following, "Global atmospheric concentrations of carbon dioxide, methane and nitrous oxide have increased markedly as a result of human activities since 1750 and now far exceed pre-industrial value determined from ice core data spanning many thousand years in the past. The global increase in carbon dioxide concentration is due primarily to fossil fuel use and land cover change while those of methane and nitrous oxide are primarily due to agriculture." This conclusion provides vital evidence about the role of human interference in climate change and reaffirms the role of international cooperation and understanding.

Research studies in climate science have received enormous impetus all over the world during the last few decades which witnessed rapid advancement in satellite-based data collection and monitoring programs. Concurrently, there has been phenomenal development in climate system models with the induction of high performance computing platforms and improved understanding of physical processes. In the year 1980, the WMO established a mechanism involving a group of experts to deal with research activities in the climate science at international level, known as the World Climate Research Program (WCRP) as an important component of the World Climate Program (WCP).The IPCC and WCRP are playing major role at global and intergovernmental levels in coordinating and guiding research and development activities in the area of climate science.

Another very important program to promote basic science of the Earth system which has made many significant contributions is the International Geosphere Biosphere Program (IGBP). The IGBP coordinated by the International Council of Science (ICSU) was launched in 1986 to look at the total Earth system, the changes that are occurring, and the manner in which changes are influenced by human actions. Findings of the scientific research under IGBP, WCRP, and the recommendations of the IPCC are therefore considered and adopted by large community of scientists and policy makers while evaluating the potential regional impacts of changing climate.

The guidelines emerging out of these recommendations thus provide the benchmark for policy makers as regards to adaptation and mitigation strategy and also provides a working paper for intercomparison and consultations among the nations. These will help in formulating the sustainable development goals as applicable to different regions while appreciating the global dimensions of the problem.

1.5 Indian National Initiative on Climate Change

With high seasonal variability of monsoons and dependence of large sector of agriculture and allied industries on its performance, the climate in India has always been a key factor in the socioeconomic well-being of the country. Recent surge in industrial and agricultural growth have generated enhanced demand and pressure on all forms of natural resources of the country. With increasing population, urbanization, and rising standards of living, India may slowly and steadily march towards the position of a major emitter of the GHG_s along with other emerging economies of South Asia.

However, at current level of GHG emission the contribution of India is far below the levels of developed industrial nations including China. The emission statistics at present may appear very pacifying considering the per capita consumption and emission data based on current level of development. The statistics based on per capita emission are generally used by policy makers while dealing with international negotiations and bargains relating to the economic and political aspects of GW. However, the increase in overall GHG emission cannot be ignored and argued for long under the protective umbrella of per capita emission as a basis. For emerging economies and developing societies, the argument based on equity and justice which envisages equality in demand and consumption provides a strong point of argument to create level field on the bargain table. However, for long period sustainable development strategy it is essential to incorporate environmental concerns in the overall developmental process at the planning and execution level.

In a wider context, it is important that all sincere and concrete measures on GHG emission control should be the ultimate objectives for all nations of the world irrespective of their economic status and regional priorities and differences. It is also necessary that the nations of the world must act as equal partners in a global framework with sustainable and suitable country norms for an overall protection to the planet Earth. The emerging economies of the world including India can take a leading role in this respect. It is not always necessary and expected that all aspects of GW may have direct and uniform environmental and socioeconomic impacts on all parts of the world. In the meantime, it must be realized that the impacts of GW on different sectors of social development need to be clearly delineated and studied as a prerequisite to relevant policy formulation and implementations.

Scientific studies in greater detail along with clear identification of vulnerability and impact analysis are the immediate requirements at national level for appropriate policy formulations and projecting the development goals. India has recently started playing an active role in international negotiations and implementing projects aimed at reduction of greenhouse gases to acceptable levels while retaining the accelerated pace of economic development. With this in mind the National Action Plan on Climate Change (NAPCC-2008) is formulated by the government of India to promote interconnected components such as understanding of climate change, adaptation, mitigation, energy efficiency, and natural resource conservation under the umbrella of a sustainable eco-friendly national development strategy.

The NAPCC provides a broad framework of strategies and actions under eight major national missions. These are National solar mission, National mission on enhanced energy efficiency, National mission on sustainable habitat, National water mission, National mission for sustainable Himalayan ecosystem, National mission for green India, National mission for sustainable agriculture, and National mission on strategic knowledge for climate change. India is an active member of the IPCC, WCRP, and IGBP since their inception with many experts from the country having made significant contributions in different capacities (including the first author as a member of all the three bodies for many years). In addition, India became an active member of the UNFCCC by signing on the10th June 1992 and further ratifying on the 1st November 1993 and also participating in the Conference of the Parties (COP) and hosting some of their thematic meetings within the country, for example, the hosting of COP-8 at New Delhi in the year 2002. Following are some examples of India's firm commitment and involvement in the climate change efforts for the last three decades.

The report of the first and opening session of the IPCC held in Geneva, 9–11 November 1988 (WMO/UNEP-TD No. 267, Annex III, pp. 3–4, 1988) documents the official Indian commitment to climate change as follows "India shares with the world community the concern and alarm about the possibilities of accelerated build-up of carbon dioxide and other greenhouse gases in the atmosphere. Wide ranging changes in climate would mean changes in monsoon, in economic activities in the entire country and would have long ranging effects in coastal areas." The actions and programs recognized and initiated by India as of the year 1988 are stated in the above document as follows, "India has already launched a big environmental protection program with afforestation, pollution control and impact assessment of human activities." As a follow-up on the above-mentioned commitment, India in the last quarter century has emerged as an active participant to the global efforts on climate change issues with positive actions at home like the NAPCC-2008 which is systematically being extended to state level.

The country had a strong observation, monitoring, and research base in the area of climate sciences since the inception of India Meteorological Department (IMD) in the year 1875. The creation of an independent ministry under the government of India, named as the Ministry of Earth Science (MoES), about a decade ago is a clear indication of the recognition and new thrust being given to the subject of the Earth system science. The seriousness in recognizing and handling of climate change-related policy matters at the political and official level is reflected in recent decision of the government of India to add climate change word in the name of the Ministry of Environment and Forests. The central ministry is now called the Ministry of Environment, Forests and Climate Change (MOEFC). In addition to the initiatives of the central and state governments, many Non-Governmental Organizations (NGOs) are actively contributing to specific projects and general awareness programs relating to clean and green environment at various levels with direct bearing on climate change issue specifically targeting the most vulnerable regions.

1.6 Climate Science: Ancient Ideas and Historical Perspective

A brief introduction to historical background of the development of scientific knowledge on weather and climate is essential before embarking on the technicalities of modern climate data collection system and introduction to modeling of climate. The knowledge base on the subject started from the early periods of ideas and concepts and lingered on for a long time till the launch of the modern age of multidisciplinary science approach. Almost contemporary to theoretical developments, a strong base on methods and techniques of observations have emerged and maintained its pace with all-round engineering and scientific innovations. One of the reasons for faster developments of technology for observational needs of earth system science in recent times has been the requirements of appropriate sensors and techniques for collection of sufficient and relevant data. These data are essential to establish and verify theoretical concepts and develop realistic numerical models for understanding and prediction of climate.

Before embarking on a comprehensive discussion about the developments of modern scientific era, it may be informative and interesting to have a quick glance at the ancient thoughts and beliefs on weather and climate. Going back to the ancient periods, it is generally believed that India is the first country to have developed the scientific basis of meteorology (a core science discipline of climate) along with the subjects of religion, philosophy, mathematics, astronomy, and languages. The

Upanishads (~3000 BC), the treatise on ancient knowledge, mention the subjects relating to the processes of cloud formation, rain, and the seasonal cycle caused by the movement of the Earth around the Sun.

The importance of the Sun as sole provider of energy to the Earth and the changes in solar energy being the fundamental cause of time and space variability of weather and rain were well known to ancient Indians. The famous sentence "आदित्यात् जायते वृष्टिः। वृष्टिः धान्यम् ततः प्रजा ।" from the ancient wisdom (Rigveda) in original Sanskrit language to this effect has found a prominent place as the official emblem of the India Meteorological Department. Once translated in English it reads as "The sun creates the rain, rains the grains, and thus the subjects." The Sun, Planets, and the Earth in particular with its components: soil, vegetation, animals, air, water, and fire have been the subjects of worship and prayers in some form or other under different faiths and beliefs in India and perhaps in many other parts of the world. Significance and purity of all these earthly components is emphasized in daily prayers and extended as sacred rituals to all Hindu religious ceremonies.

A critical examination of the basic contents of these traditions suggests a clear understanding about the significance attached to the Earth system components and special reverence and importance attached to the Solar system. In different parts of India, the rainy season finds highly significant place in the life of rural folks with celebrations galore studded with folklores, songs, dances, festivals, rituals, art, and paintings. Behavior of animals, migration and breeding pattern of birds, color of the sky, and flowering of plants are some of the events linked to the change of weather, seasons, and the arrival of rains. The farmers all over the world have been using thumb rules to decide the occurrence of weather events for sowing, watering, harvesting, and drying the farm products for centuries based on their experience, observations, and sometimes mere traditions and belief. The most prominent among the early scholars of India often quoted in context of in-depth understanding of the subject and noteworthy contributions include many legendry figures such as Manu, Bhrigu, Panini, Chanakya, Bhaskaracharya, Varahamihira, Aryabhatta, and Kalidas.

The centers of learning such as Nalanda and Taxila have eminent scholars from far and wide who attempted to study some aspects of weather, rains, clouds, winds, and many other fascinating phenomena in the atmosphere like the lightening, rainbow, twilight, and color of the sky. Chanakya (400 BC) in his Kautilya Arthashastra (economics) describes the system of quantitative measurement and record keeping of rainfall for the purpose of classifying land wetness as a yard stick for land revenue collection during the Maurya dynasty. The Greek philosopher Megasthenes, an ambassador to the court of Maurya kings around 400–300 BC, has provided descriptions of rains, winds, trees, plants, and animals suggesting a monsoonal climate over India. There are descriptions of winds over the Ocean surface and many accounts of different ways to navigate through monsoon winds across the Indian Ocean in the Greek sailor's accounts dating back to around 100 BC. The mariner's log books suggest that many merchants meticulously followed these paths and benefitted by their guidance. Around AD 400, Fa Hsien, a Buddhist scholar who visited India from China describes the strong monsoon winds and rain which he observed while sailing along east coast of the Indian subcontinent.

The classical work of well-known Indian astronomer Varahamihira (500 AD) provides many ideas relating to the subjects of atmosphere and the weather. Famous Sanskrit scholar and poet Kalidas in his classical treatise Meghdoot (700 AD) describes the date of onset of monsoon over central India (Ujjain, Madhya Pradesh) as the first day of *Ashad*, the Indian calendar month (approximately the first week of July) which is close to the normal date of arrival of monsoon there. He also descriptively traces the path of monsoon clouds in the sky as the wind-driven messengers. His description of the monsoon flow and the movement of clouds in the sky over the Indian region have been very much similar to the modern observations.

It is rather surprising and also disappointing that even after many centuries of logical understanding of weather, climate, environment, and related events and the significance attached to them; not much of systematic documentation of information or compilation of data is available in ancient Indian literature. The state of affairs is not much different from other ancient civilizations including the Greeks and Chinese. Therefore, no chronological sequence of development of knowledge can be traced or formulations can be made out of those understandings and observations. As a result, it is not possible to apply the ancient wisdom to present day computations for understanding and prediction of weather and climate.

Globally speaking; Thales of Miletus (Greek, 624–546 BC) was the first to discuss the water cycle akin to the hydrological cycle known today. Aristotle (Greek, 384–322 BC) a student of well-known philosopher Socrates (Greek, 469–399 BC) compiled the first treatise in meteorology, "The Meteorologia." He postulated and provided conceptual explanations on various weather phenomena which remained the guiding principles on the subject for almost 2000 years though many of his descriptions could not stand scrutiny based on modern observations. Functioning of the climate system in his own words is described as follows: "…all the affections we may call common to air and water, and the kinds and parts of the earth and the affections of its parts." One of the greatest achievements of this period is the apt description of the hydrological cycle in the following original words. "Now the Sun, moving as it does, sets up process of change and becoming and decay, and by its agency the finest and sweetest water is every day carried up and is dissolved into vapour and rises to the upper region, where it is condensed again by the cold and so return to the Earth."

His contemporary Theophrastus (Greek, 371–287 BC) compiled a book on weather forecasting named "The Book of Signs" mostly based on subjective observations and prevailing notions. The well-known principle of Archimedes (Greek, 287–212 BC) is a basic concept to the study of atmospheric flows and buoyancy vital to explain the formation of clouds and the rain. In the year 25 AD, Pomponius Mela a geographer in Roman Empire formalized the importance of the climate zones. Wang Chang (Chinese, 27–97 AD) had described the process of rain formation through the clouds and their formation. The Greek astronomer and geographer Claudius Ptolemy (85–160 AD) and many of his followers believed that the constellations of planets (the Sun and the Moon included) affected the atmosphere. Therefore, the astronomical conjunctions of stellar objects and movements of planets and stars were seen as the ways to formulate and pronounce the long-term prediction of weather and climate. The Chinese have many ancient manuscripts, traveler's logs, stories, and paintings about the seasons and weather events since about 200 AD and so do the Greeks.

Al Dinawari (Persian, 828–89 AD) wrote a book entitled Kitab-Al-Nabat (Book of plants) which describes the application of meteorology to agriculture in which he deals with many weather phenomena and their influence on vegetation. In another book entitled *The Book of Balance of Wisdom* published in the year 1121 AD by famous Islamic astronomer Al-Khazini (Persian-Greek, eleventh to twelfth century AD), there is a scientific description of the hydrostatic balance in the atmosphere. Johannes Kepler the famous astronomer (German, 1571–1630 AD) was the first to present a scientific treatise on ice crystals in addition to his famous work in astronomy. In the year 1636, Edmund Halley published a treatise on the Indian Summer Monsoon, which he attributed to a seasonal reversal of winds due to the differential heating of the Asian land mass and the Indian Ocean.

It is interesting to note that the developments in the modern science in Europe and later in the United States have vigorously followed the earlier attempts of natural philosophers and scientists such as Plato (Greek, 428–348 BC) and Aristotle (Greek, 384–322 BC), the contemporaries of the famous philosopher Socrates (Greek, 469–399 BC) to conceptualize the natural processes on the Earth. This was followed by the legacy of the legendry figures, namely, Archimedes (Greek, 287–212 BC), Galileo Galilei (Italian, 1564–1642 AD), Johannes Kepler (German, 1571–1630 AD), Isaac Newton (British, 1643–1727 AD), Edmond Halley (British, 1656–1742 AD), George Hadley (British, 1685–1768 AD), Alexander Von Humboldt (Prussian, 1769–1859 AD), William Ferrel (American, 1817–1891 AD), and many others in natural sciences.

On the observational front, the first design of a Hygrometer, the instrument to measure humidity of air was discovered by Nicholas Cusa (German, 1564–1642 AD). Galileo Galilei invented an early thermometer and Evangelista Torricelli (Italian, 1604–1647 AD) invented the Barometer to measure the atmospheric pressure. The famous painter of Mona Lisa fame Leonardo da Vinci (Italian, 1452–1519 AD) is credited with the original idea on measurement of wind speed and direction as well as the atmospheric humidity. In the year 1667 AD, Robert Hooke (British, 1635–1703 AD) built another type of anemometer called a pressure plate anemometer.

Many physicists in the modern Europe worked on the quantification of heat exchanges in physical processes and chemical reactions, with focus on the measurement and definition of heat energy, thus initiating the initial concept of thermometry. Daniel Gabriel Fahrenheit (Polis, 1686–1736 AD) was the first to develop the reliable mercury in glass thermometer with graduations called Fahrenheit scale (°F), later named after him. Anders Celsius (Swedish, 1701–1744 AD), an astronomer suggested modern thermometer scale with graduations in the range defined by the freezing and boiling points of pure water at normal atmospheric pressure called the Centigrade scale, (°C) later named after him as the Celsius scale.

Lord Kelvin/William Thomson (Irish, 1824–1907 AD) determined the correct value of absolute temperature, a state at which no further work can be performed with transfer of heat to mechanical energy, a state of absolute rest according the kinetic theory of molecules. The temperature scale starting at −273.15 °C, the absolute zero, is named after him as degree Kelvin (°K). The seventeenth to nineteenth century Europe has seen extensive research in the subject of thermodynamics with a large number of famous scientists associated with the discovery of the basic laws of physics relating to heat and energy transfer and the kinetic theory of gases. Readers may find it interesting to learn more about the life and discoveries of the scientists during that period. Some of the suggested readings may relate to Robert Boyle (Irish, 1627–1691 AD), Jacques Charles (French, 1746–1823 AD), James Prescott Joule (British, 1818–1889 AD), Rudolf Clausius (German, 1822–1888 AD), and Benoit Paul Clapeyron (French, 1799–1864 AD).

As stated earlier, the rainfall amount was being measured in India almost about 2400 years ago. The credit for quantitative measurement and record keeping of rain amount is given to king Sejong of Korea (1397–1450 AD). The tipping bucket rain gauge was invented by Christopher Wren (British, 1632–1723 AD) in Europe around 1661 AD. The oldest climate data on surface temperature and rainfall with daily observations started developing in Europe with some important stations maintaining records as far back as seventeenth century. In the year 1654 AD, the Italian prince Duke Ferdinand II of Tuscany sponsored the first weather observing stations at ten places in the part of Europe. The data collected from these stations was sent to Academia del Cimento in Florence (Italy) at regular intervals. There are records of central England temperatures since 1659 AD and precipitation of Kew, England since 1697 AD and also the precipitation of Paris since 1680 AD (Brazdil et al. 2005). Benjamin Franklin (American, 1706–1790 AD), the versatile scientist and a Statesman of America, was the first to start the practice of rainfall (rain depth in inches) measurement in the United States.

Astrometeorological predictions for agriculture and general well-being of the masses became a practice in Europe with further refinement in Asia and the Arab world almost 2000 years back. Establishment of this tradition prompted the record keeping of meteorological observations in many parts of the globe as supplementary information to extensive records relating to the movement of extraterrestrial objects in the sky. With the advent of printing technology around the seventeenth century, the astronomical calendars with ephemerides on weather became the best sellers in many parts of the world; the practice which continues even today. In India, the *Panchang*, an astronomical calendar printed every year and issued on the auspicious day of the start of local New Year (*Vikram Samvat*) specifically mentions the likely performance of the weather events, such as monsoon and winter precipitation during the following year. The practice in general commands large acceptance and faith without demanding factual verifications or scrutiny at the end of the year as in case of official weather predictions.

An all-round development in multidisciplinary science disciplines during the nineteenth and early twentieth century have successfully created a solid backbone of climatology as a science discipline of pure and applied nature. The physical understanding of the solar system and geophysical details of the planet earth became well established. Almost concurrently, the basics of gravity, energy, radiation, thermodynamics, fluid dynamics, mathematics of computation as well as the physics of precipitation and clouds remained at the center stage of natural science research.

Big boost to the study of thermodynamics of water substance came with the invention of steam engines (James Watt Scottish, 1736–1819), atmospheric fluid dynamics from the invention of aircraft (Wright brothers: Orville, 1871–1948, Wilbur, 1867–1912) and modern rocketry (Robert H. Goddard, American, 1882–1945). Though the scientific expeditions and observational and theoretical studies of the Oceans have an equally exciting history, in context of modern climate science the study of the Ocean surface characteristics (waves, winds, temperature, etc.) received special impetus due to increased marine navigation. On technological front, the data transmission and coding system developed due to the invention of wireless telegraphy first demonstrated by Guglielmo Marconi (Italian, 1874–1947 AD) in 1897 AD. Many formulations and concepts of fluid dynamics were applied to the problems of the Earth system science with unified goal of using them to the problems of weather prediction. In fact, a new discipline of computational geophysical fluid dynamics emerged around mid-twentieth century in the United States involving many eminent mathematicians and computational experts, initially at the Princeton University. The discovery of depleting O_3 in the stratosphere over Antarctica later recognized as the ozone hole along with the deteriorating air quality in the modern cosmopolitan cities and also the injection of pollutants in upper atmosphere by the jet aircraft and rockets put together have given big boost to the subject of atmospheric chemistry and allied sciences.

Last but not the least is the development of technology during the recent times focusing at the need for energy efficiency and invention of alternate and renewable energy sources. The motivation for new outlook towards renewable energy sources is essentially related to the requirement of controls to be imposed on the emission of GHGs, an issue intimately related to climate change. In fact, the emphasis on energy efficiency, conservation, and search for environment-friendly sources has been mainly propelled ahead because of the compulsions of the climate change.

A large number of these important discoveries of modern science were simultaneously followed by developments of technology seeking for the modern tools of observations, data collection, transmission, and computation. Application of remote sensing satellites as data collection platforms and many dedicated satellites for weather monitoring as well as large memory-high speed computers for simulation and forecast have revolutionized the arena of earth system observations and computation. Studies on microphysics of clouds, rain, and other related weather processes received intensive support during the last century to explore the possibility of weather modification as a means of mitigating the effect of droughts by the artificial rain making.

The scientific experiments aimed at enhancing rainfall over water scarcity regions and suppression of the furry of cyclones and to reduce the damage caused by hailstones known as cloud seeding are being conducted in many countries. The feasibility, effectiveness, and acceptable cost-benefit ratios of these experiments are being widely debated and need extensive scientific evaluation and monitoring. Notwithstanding these issues, it is certain that these experiments in general have helped in generating data bases on physical processes within the clouds, an essential requirement for cloud microphysical studies which also improve the inputs of high resolution weather prediction models.

As an example of new and innovative ideas on the observational front, it is interesting to recall the case of microwave sensors such as RaDAR, originally conceived and designed to detect aircraft and objects in the sky. These devices are now being extensively used as ground and satellite-based sensors for the study of clouds and hydrometeors (rain, snow, hail, etc.) by collecting quantitative data on the microphysical processes within the clouds. Satellite-borne RaDAR observations are extensively used for quantitative estimation of rainfall on the ground as well as monitoring of the mesoscale intense weather systems. The theoretical concepts and discoveries involving the basic science principles such as absorption, emission and scattering of radiation, Doppler-effect, microwaves, LASER, infrared and visible spectroscopy, photochemical reactions, photosynthesis, photoelectric effect, acoustic soundings, mass spectrometry, radio activity, and thermal imaging are all finding an important place in the development of modern scientific observation tools for weather and climate.

Most of these early developments of modern meteorology took place in Europe; however, almost contemporarily they were brought into the Asian monsoon region to generate data and to create a strong weather and climate science base. In India, during the British rule, the importance of monsoon prediction and cyclone monitoring have received priority due to socioeconomic and political reasons of governance particularly during the periods of frequent droughts/floods and devastating cyclones. A unified administrative control and strong legacy of scientific knowledge on the subject within the country have helped this transformation to happen uninterruptedly and over a wider area.

From the science point of view, the tropical regions (a belt of approximately 30°N–30°S from the equator) are highly significant and have received the attention of the world scientific community during the recent decades. The tropics have an excess balance of heat received from the sun throughout the year with largest area under oceans to store and provide flux of heat and moisture to rest of the globe. The persistent surface wind over the tropics, appropriately named as trade winds have helped and fascinated the mariners and the users of the sea trade routes for centuries. During the periods prior to the establishment of formal weather services many devastating tales of cyclones of the tropical Oceans have remained in mariners log books, often read as the triumph of valor off shore helplessly. The spirit of enquiry and scientific investigation of last few centuries have led to many expeditions to polar regions, glaciers, and areas of frozen seas. These expeditions had many setbacks and tragedies but the human endeavor continued with numerous success stories leading to significant scientific discoveries. This has helped climate science in many ways by improving the understanding of climate system components and their mutual interactions and roles.

During the last about two centuries, the weather observations have steadily increased in spatial coverage, quality, and increasing number of essential parameters to be measured. Recent emphasis on observations over the tropical Oceans has been primarily motivated by the scientific, strategic, and economic importance of the region. Practically, all over the world the observation programs relating to climate science have been funded and encouraged by the governments of the respective countries through public institutions. This has helped in the establishment of global observation network and consistency and sustainability in operation and practices.

The persistent encouragement and initiatives have helped in developing better global coordination, continuity, and uniformity in the development of strong data base with solid foundation. In the initial phases of establishment of the global observational network, generous support from the rulers, local governments, and the world bodies such as the United Nations have encouraged the development of an integrated global system of observation for weather records and prediction.

The encouragement received by the observational programs of significance to weather and climate all over the world has been based on the following main premises.

i. The weather forecasts have been realized to be very essential for improving the prospects of food availability and crop production.
ii. The weather services are recognized to be essential for helping the sailors to survive and navigate across the open seas.
iii. The weather is a critical parameter in the area of multipurpose aviation and allied services. With an accelerated rate of development in navigation through greater heights in the atmosphere, the requirements of instant information, long period data, and route weather forecasts have enormously increased with continuous growth.
iv. Weather services are considered to be vital for successful intercontinental trade relations and military operations during war and peace.
v. The weather information is very useful for the governments to issue advance warnings to public and to manage and reduce the risk and loss due to natural disasters such as drought, floods, and high impact extreme weather events.
vi. The long period data on weather parameters that create a climate base are extremely useful in planning and executing the developmental activities and natural resource management programs.
vii. The weather forecasts and weather-related information has now become an integral part of day-to-day life of practically every one. The customized information on climate is therefore playing an important role in planning process for individuals and communities.

1.7 Observations and Data for Climate Change Study

For meaningful study and analysis of climate change, the climatologists need long period of continuous data on parameters such as rainfall, temperature, pressure, wind, and humidity near the earth's surface. In addition to surface meteorological data recorded continuously over fixed interval of time, other relevant information from the Oceans, biosphere, cryosphere, and land surface are equally important for complete study, description, and future projection of climate. Environment of the earth being a three-dimensional envelope also requires data at vertical levels in the atmosphere as well as the depths of the Oceans for a comprehensive description of climate and its change.

In addition to quantitative data from the modern instrumental observations, it is also essential that the relevant information of reliable quality with sufficient spatial and temporal resolution is collected for the periods prior to the instrumental records. Information for the past can be extracted from many natural indicators which have preserved the footprints of past changes in their repository as proxy data source. The data sets generated using the proxies are very useful to extrapolate climate information backward in time for creating qualitative/quantitative knowledge base on climate for much longer periods. These data may be useful for the study of variability and changes in climate on longer time frames of decades, centuries, and even beyond. Some of the important natural sources of proxy data are historical accounts, archeological findings, lake pollen deposits, sedimentary rock deposits, cave speleothems, soil forms, ice cores, tree ring records, and the marine sediments deposited below coastal waters and the Ocean floors. The branch of science dealing with the study of past climates based on the information extracted from proxy sources is known as Paleoclimatology.

For a complete picture of climate to emerge, a lot more of data are being acquired over and above the data on the basic meteorological parameters mentioned above. These include the data on solar radiation, thermal radiation emitted by the earth's surface, concentration and distribution of atmospheric GHGs, aerosols, dust, and other particulate matter including other minor gaseous constituents and their vertical profiles. Initial and boundary conditions truly representing the processes within and at the interfaces between various components of the climate system are generally denoted as representative parameters over predetermined three-dimensional grid boxes (latitude-longitude and vertical layers) all around the globe at fixed time intervals to carry out model simulations. Therefore, continuous monitoring, analysis, and assimilation of data for description, modeling, and prediction are important components of the modern developments in the science of climate change.

As a result of all-round appreciation of weather-related data and services and their global dimension, it was convenient to organize international cooperation in meteorology as early as 1873 AD in the form of the First International Meteorological Congress (Vienna, September 1873). A historic decision of the congress was the establishment of non-governmental International Meteorological Organization (IMO) the predecessor to the current World Meteorological Organization (WMO). The WMO came into being on 23 March 1950 as a specialized agency of the United Nations and have the current membership representing almost all nations of the world. During the first decade of its existence, the WMO established a landmark program for observations called the World Weather Watch (WWW). This initiative included three major components, namely, the Global Observing System (GOS), the Global Data Processing System (GDPS), and the Global Telecommunication

System (GTS) implemented practically over the entire global land mass in coordination with the weather services of all the member countries.

Modern meteorological observations in the Indian region began with the amateur observations of atmospheric pressure and temperature by the East India company towards the beginning of nineteenth century. The earliest surface meteorological observations for Madras (now, Chennai) are available since 1813 AD which contain continuous record of monthly rainfall though the continuous surface temperature record started much later. The second observatory in the country started functioning since 1817 AD at Bombay (now, Mumbai). For Mumbai continuous record of rainfall amounts from 1817 to 1841 is available only for monsoon months; and thereafter, the annual rainfall and other surface meteorological parameters have continuous record.

In Bengal, the first observatory was set up in Calcutta (now, Kolkata) in the year 1829 AD with rainfall and temperature initially being collected by the Asiatic Society of Bengal. Subsequently, surface observatories were established at Bangalore (now, Bengaluru), Shimoga, Trivandrum (now, Thiruvananthapuram), Hyderabad, Patna, Allahabad, Shimla, and Mukteshwar. During the early years, observations were mainly of rainfall, temperature, and occasionally atmospheric pressure. Many of these early records up to about the year 1860 AD have serious limitations owing to disparities in the exposure of instruments, site change, nonstandard times, and observing practices and inadequate calibration.

By the middle of the nineteenth century, meteorological observations were extended to several other parts of the country including the parts of the North-West Frontier Province (part of which are now in Pakistan). The observational network over the country was consolidated under a single agency with the establishment of IMD in the year 1875 AD thus providing big boost to weather observations and services. This has helped in the establishment of a unified forecasting system and systematic compilation and record keeping for climate data. Currently, data of climatic significance for surface and upper air are collected and monitored by IMD with high density of stations over many land areas and adjoining oceanic regions. Currently, the IMD is operating about 675 Automatic Weather Stations (AWS) including 127 Agro AWS, 1330 Automated Rain Gauges (ARG), and 42 upper air stations along with a large number of surface observatories for routine measurements. IMD has also expanded its Doppler Weather Radar Network from 14 in 2012 to 24 in 2016. As far as the daily total rainfall measurements are concerned, various government departments are augmenting the IMD data for spatial homogeneity and applications for hydrological and water resource purposes. This involves data collection and processing for thousands of stations across the country to prepare authentic rainfall climatology.

The data available from these sources is processed, archived, and routinely used in operational weather forecasts and climate studies. The data platforms today constitute a composite network of high density land-based observing stations, dedicated satellites, oceanographic buoys, and platforms with an objective of attaining global consistency. The climate data involving many parameters related to the integrated earth system processes is routinely observed and distributed by the government departments all over the globe. However, information from other sectors is also integrated into the data archives wherever authentic data is available. Additional data from international sources mostly in the form of satellite-derived products and compilations from special field campaigns over land and Oceans are assimilated routinely in the weather and climate prediction system and records are maintained.

Many researchers and operational meteorologists are today using the processed, interpolated, and assimilated data sets on global scale which are freely available to genuine users, commonly termed as the reanalysis data sets (NCEP-NCAR, USA). The reanalysis is a systematic approach to produce global data sets for climate monitoring and research under which millions of periodically updated observations from all sources are used in a standard assimilation scheme and modeled to generate grid point data. Global data sets with many variables and consistent spatial and temporal resolution are freely available on climate timescale of more than three decades with variable grid dimensions starting with coarse grids of 2.5-2.5 degree to high resolutions of few kilometers in space and few hours to days in time domain.

While using these data sets over complex terrains like Himalaya, it needs to be remembered that the observational constraints over the region may limit the applicability of data to smaller spatial scales. In fact, under certain situations such as the diagnostic and prognostic analysis of extreme weather events which evolve within a timescale of few hours, the model generated data may produce misleading results. This may specifically happen when the retrieval techniques of satellite observed data over the variable complex terrain are not thoroughly calibrated with the ground truth, which is generally not feasible over all geographical sectors and the time resolution is much longer than the scales of few hours.

Additionally, data pertaining to space-borne measurements, particularly the satellite observations across the globe are freely available for research and application needs. In addition to dedicated indigenous Indian satellite missions such as INSAT, KALPANA, Megha Tropique, and OCEANSAT for the Indian climatologists, there are many other global sources available for data augmentation. Presently, one can have an access to relevant data archived through the satellites of other countries like MODIS, CALIPSO, TRMM, and many more.

With this background of rapid development in observational platforms and availability of more than a century of reliable high resolution surface and upper air data over most part of the country India today is in an advantageous position in regard to the issues related to the factual information on climate change. The Indian climatologists can now authentically assess the climate trends over a century for the country as a whole or its subdivisions and contribute to global climate analysis, research, and prediction efforts at equal footing. For the Indian region, analysis of temperature data over a large number of stations since 1901 till 2015 suggests a linear trend of 0.59 °C/100 years for all-India annual mean surface temperatures. Similar trends are reported for smaller subdivisions with variations during different month and seasons.

An All-India annual surface temperature plot based on the Indian Institute of Tropical Meteorology (IITM) homogeneous Indian monthly surface temperature data for the period 1901–2012 is presented in Fig. 1.5; however, the data series up

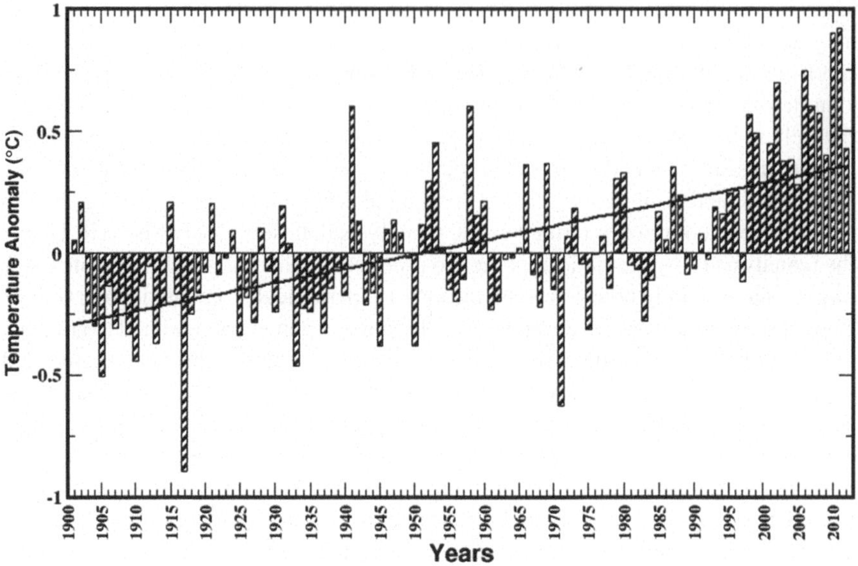

Fig. 1.5 All-India annual surface temperature anomalies (departure from the mean of entire period) in °C for the period 1901–2012 indicating the trend line

to 2007 has been presented elsewhere (Srivastava et al. 2009; Kothawale et al. 2010). It is clearly seen in the analysis of data over the Indian region that the last few decades have witnessed many anomalously warm years (Climate Diagnostic Bulletins, http://www.imd.gov.in) and an enhanced increasing trend in surface temperature. These data are comparable with the trends of global mean temperatures reported by IPCC and other international research and service groups.

The rainfall in general is a resultant product of many complex atmospheric and oceanic processes which involve variety of direct and indirect influences and scales of motions. Therefore, it is extremely difficult to detect the impact of increasing temperatures on rainfall by examining the rainfall trends over a geographical area even on seasonal or annual scale. All-India rainfall series prepared by the IITM, Pune (http://www.tropmet.res.in) based on data since 1871 till the monsoon of 2016 from a large number of homogeneously distributed area weighted rain gauge stations (306) for the monsoon season (June to September) is presented in Fig. 1.6. The rainfall amounts are over an area covering the entire monsoon region for the period June to September (monsoon season), which in average contributes about 85% to the total rainfall. This climatic data set along with the data warehouse of IMD provides very useful tool for the analysis of trend and variability in annual and seasonal rainfall over India and also for smaller meteorological subdivisions even up to district level (http://www.imd.gov.in).

The analysis of this rainfall series does not depict any statistically significant trend but displays large temporal variability with clear wet and dry epochs on

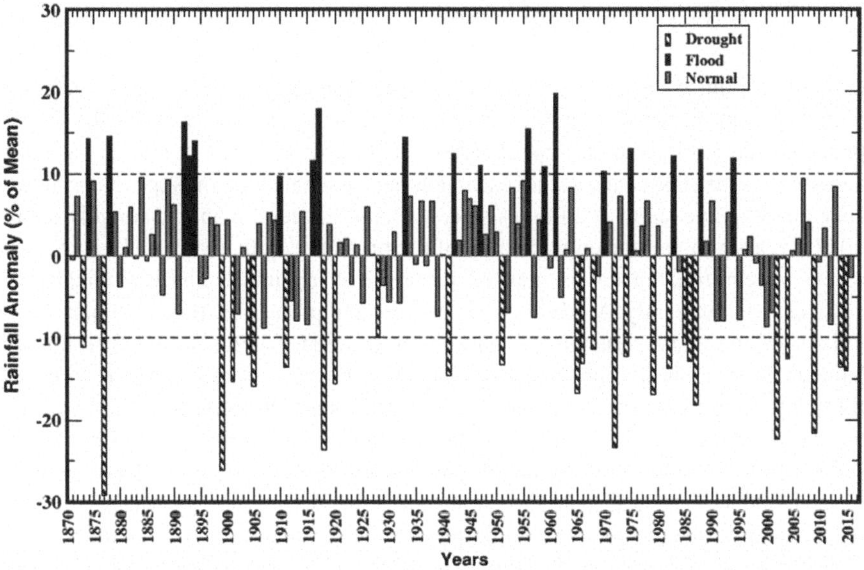

Fig. 1.6 Variation of all-India summer monsoon rainfall anomalies (departure from mean expressed as percentage of mean) for the period 1871–2016, based on the area weighted average of 306 stations; *horizontal dashed lines* indicate one standard deviation above and below the mean

decadal scale extending from three to four decades as well as a strong signal of interannual variability. These fluctuations are embedded on an otherwise stable long-term monsoon rainfall behavior displayed by the data of last about one and a half century. With an estimate using all possible rainfall records, it is worked out that the total rainfall over the country for monsoon season (June–September) is approximately 85 cm with a standard deviation of 8.5 cm (10% of the mean). Largest variability of 30% is seen over the north-west India and the smallest variability of 8% over the northeast India. For details on the rainfall variability and behavior and their characteristics over the country as a whole as well as the meteorologically homogenous subdivisions, the readers may refer to a book by Pant and Rupa Kumar (1997) and other recent research publications on the subjects including the official reports of IMD.

The winter rains comprising the rains due to western disturbances over NW India and the NE monsoon rains over the eastern peninsular India are also random in behavior depicting an insignificant trend and a large interannual variability (Sontakke 1993). Other months of pre- and post-monsoon consist of spatially inhomogeneous rain events with thunderstorms with localized intense rain occurrences; thus, it is not appropriate to define them in terms of areal averages. However, there are many studies suggesting an increase in the extreme weather events during the last few decades associated with these convective activities which may have an impact on total rainfall. Very small and statistically insignificant decreasing trend in the all-India average monsoon rainfall as seen in Fig. 1.6 is an indication of stable behavior of monsoon rainfall over the country on long-term basis.

1.8 Climate Change and Society

Climatic conditions over the surface of the earth for a period of time in association with their variability and change constitute a significant factor in deciding the social, political, economic, and cultural changes which have occurred in human societies and their dominance of the planet earth. Humans have had the most profound influence on the planet earth at all times since the emergence of mankind roughly about 10,000 years ago. For the entire history of life on planet earth, mankind has the most recent appearance but has influenced its physical, chemical, and biological processes most abundantly. All through the historic and prehistoric times, humans have influenced the planet, changing conditions of the planet in turn influenced human habitats, livelihood, and processes of adaptation to changes that continue to occur till today. Along with these changes, the techniques and magnitude of exploitation, mutual dependence, and dominance have evolved with time assimilating subtle refinements and distortions. So far the processes of development remained confined to sustainable existence with principles of adjustments and sharing of natural resources the concerns of society at large have been manageable and adaptable.

Emergence of a rapidly developing society and a new social order since last about half a century have acquired the dimensions of significant manipulations of environmental norms for selfish motives, impacting on climate system and creating constraints for normal adaptation strategy. This influence in last few decades have acquired such a dimension that multifaceted impacts on ecosystem and society have started emerging in different forms and quite often the reactions of society to these changes may create unpredictable adverse social problems. It is certain that the impacts which the parameters of climate change will incorporate into the availability and quality of all living and nonliving natural resources will make the proposition of survival on earth much more complex than expected. Prevailing social systems and practices may not be adequate for sustainable and healthy growth of human societies to keep pace with fast growth and high level of consumption patterns in the background of overall environmental impacts.

Though the curiosity to seek answers to these vital questions may appear trivial, the stakes in ignoring them are quite large and need urgent attention. Climate change is now recognized as a phenomenon that will be seen and experienced by people all over the world though the degree of awareness and perceptions may vary. The impacts of climate change and the vulnerability of ecosystem and society to these changes are neither evenly distributed within the regions nor among different communities and sectors of society. Perceptions on climate change may vary from person to person and will depend on their location, circumstances, and the level of individual comprehension.

Individual to community and then to regional level understanding and perceptions on climate change issue is likely to grow from subjective opinions to quantitative data. The rate of growth may become faster as the means to communicate, discuss, and analyze local events in larger context develop further with the spread of education and information technology. Substantial damage to natural resources coupled with loss of life and property make people aware and concerned, but in the absence of a reliable mechanism and appropriate procedures for the assessment of vulnerability, adaptation, and long-term effects, the process remains complex. In general, these issues with global ramifications are usually considered to be beyond the comprehension of an individual or a small group thus minimizing the individual participation. The communities to the best of their memory and understanding tend to recollect the natural hazards of different magnitude and duration which they experience during their life time. These include landslides, cloud bursts, mud flows, instances of pests and deseases, floods and droughts, crop failure, and loss of life and property. Therefore, they often describe these as direct manifestation of changing climate. The most popular perception about climate change which the individuals or smaller groups of people generally have in their mind are broadly related to unseasonal rains and unprecedented intense weather events having direct impact on agriculture, livestock, and other basic needs of their day-to-day life. Therefore, in order to provide useful advice to the vulnerable sections of society many organized and in-depth research studies with quantitative data and information is essential. These studies should specifically link climate change and fluctuations with food chain, hydrological cycle, and major health issues.

References

Arrhenius SA (1898) On the influence of carbonic acid in the air up on the temperature of the ground. Philos Mag 41:237–276

Brazdil R, Pfister C, Wanner H et al (2005) Historical climatology in Europe—the state of the art. Clim Change 70(3):363–430

Callendar GS (1938) The artificial production of Carbon dioxide and its influence on temperature. Q J R Meteorol Soc 64:223–237

Du MY, Kawashima S, Younemura S, Zhang XZ, Chen SB (2004) Mutual influence between human activities and climate change in the Tibetan plateau during recent years. Glob Planet Change 41:241–249

Hawkins E, Jones PD (2013) On increasing global temperatures: 75 years after callendar. Q J R Meteorol Soc 139(677):1961–1963. ISSN: 1477-870X. doi:10.1002/qj.2178

Hays JD, Imbrie J, Shackleton NJ (1976) Variations in the earth's orbit: pace makers of the ice ages. Science 194(4270):1121–1132

IPCC-WG-1 (2013) Climate change 2013. In: Stocker TF, Quin GK et al (eds) The physical science basis. Cambridge University Press, Cambridge. 1535p

Jansen E, Overpeck J, Briffa KR, Duplessy J-C, Joos F, Masson-Delmotte V, Olago D, Otto-Bliesner B, Peltier WR, Rahmstorf S, Ramesh R, Raynaud D, Rind D, Solomina O, Villalba R, Zhang D (2007) Palaeoclimate. In: Solomon S, Qin D, Manning M, Chen Z, Marquis M, Averyt KB, Tignor M, Miller HL (eds) Climate change 2007: The physical science basis. Contribution of Working Group I to the Fourth Assessment Report of the Intergovernmental Panel on Climate Change. Cambridge University Press, Cambridge

Kothawale DR, Munot AA, Krishna Kumar K (2010) Surface air temperature variability over India during 1901–2007, and its association with ENSO. Clim Res 42:89–104

Meehl GA, Washington WM (1993) South Asian summer monsoon variability in a model with doubled atmospheric carbon dioxide concentration. Science 260:1101–1104. doi:10.1126/science.260.5111.1101

Mooley DA, Pant GB (1981) Droughts in India over the last 200 years, their socioeconomic impacts and remedial measures for them. In: Wigley TML, Ingram MJ, Farmer G (eds) Climate and history: studies in past climates and their impact on man. Cambridge University Press, Cambridge, pp 465–478

Muller RA, McDonald GJF (1997) Glacial cycles and astronomical forcings. Science 277(5323):215–218

Pant GB, Rupa Kumar K (1997) Climates of South Asia. John Wiley & Sons, New York/London. 320p

IPCC (2007) Summary for policymakers. Climate change 2007. Impacts, adaptation and vulnerability. In: Parry ML, Canziani OF, Palutikof JP, van der Linden PJ, Hanson CE (eds) Contribution of working group II to the fourth assessment report of the Intergovernmental Panel on Climate Change. Cambridge University Press, Cambridge, pp 7–22

Schlesinger E, Mitchell JFB (1987) Climate model simulations of the equilibrium climate response to increased carbon dioxide. Rev Geophys 25:760–798

Sontakke NA (1993) Fluctuations in north east monsoon rainfall over India since 1871. In: Keshavmurty RN, Joshi PC (eds) Advances in tropical meteorology. Tata McGraw-Hill Ltd., New Delhi, pp 149–158

Srivastava AK, Rajeevan M, Kshirsagar SR (2009) Development of a high resolution daily gridded temperature data set (1969–2005) for the Indian region. Atmos Sci Lett 10:249–254. doi:10.1002/asl.232.

Chapter 2
The Himalaya

अस्त्युत्तरस्यां दिशि देवतात्मा हिमालयो नाम नगाधिराजः।
पूर्वापरौ तोयनिधी विगाह्य स्थितः पृथिव्या इव मानदण्डः॥

[In the north, there stands Himalaya the King of the Mountains, having a divine soul. It exists like a measuring rod of the earth, having reached the eastern and western seas]
Kalidasa—Kumarsambhava 1.1

2.1 Evolution of Himalaya and Present Geographical Setting

The earth as a planet of the solar system is estimated to have formed about 4.6 billion years ago. It has gone through many evolutionary geological and biological changes and formations during the course of its journey to the present day lively

G.B. Pant et al., *Climate Change in the Himalayas*,
DOI 10.1007/978-3-319-61654-4_2

planet. One of the most important geological events in the history of the earth during the last 100 million years is the formation of the Himalaya. German meteorologist and geophysicist Alfred Lothar Wegener (1880-1930 AD) proposed the theory of continental drift based on the principle of plate tectonics in 1912 AD to explain the mechanism of changes in the surface layers of the earth. This theory postulates that the upper crest of the earth consists of a combination of solid drifting plates which glide over or may even collide with each other. Accordingly, the geologists suggest that around 125 million years before present (mybp) the earth's crust with drifting land masses was constituted of two major continents, Laurasia and Gondwanaland separated by Tethys Sea. The Gondwanaland comprised peninsular India, Australia, Madagaskar, Africa, and South America (all of which can be seen to fit together into one mass, like the pieces of a jigsaw puzzle).

Gigantic earth movements and volcanic eruptions separated the continents from the Gondwanaland. Northward journey of Indian plate relative to Australia and Antarctica is estimated to have started around 120 mybp. It is estimated that by about 55 mybp the Indian subcontinent was close to its present position as a breakaway piece from the giant Gondwanaland. The mighty Himalayan ranges emerged due to the impact of the Indian land mass with the Asian land mass to its north, followed by a phased uplift between 50 and 40 mybp. This uplift was a result of the continuous upward push of the Indian land mass in phased manner. The most recent phase which occurred about a million years ago gave the mountain ranges its present geomorphic form. The fault lines along the impact regions are fragile and contain the belts of volcanic and seismic activity. An example of this is what was witnessed during the recent devastating earthquake across central Himalaya with its center over Nepal followed by an equally strong earthquake in the north-west Himalaya in the year 2015.

The Himalaya which literally means abode of snow is the greatest east-west oriented mountain range of the world constituting of the youngest mountain system on the Earth, still rising in its height and drifting northward at a slow pace of about a meter per century. These mountain ranges are extensively spread from west to east in a total length of about 2500 Km all along the northern boundary of the Indian subcontinent from Pakistan in the west to frontiers of Myanmar in east and have an average north-south extent of about 200 km. the geographical map of south Asia depicting Himalaya is given by Fig. 2.1. The term Himalaya includes the snow-covered mountain peaks, extensive mountain ranges rising from northern plains of the Indian subcontinent up to the Tibetan Plateau. It is characterized by vast biomass reserve, highly fragile ecosystem with many earthquake and landslide prone zones, large biodiversity, and a chain of glaciers linked to extensive network of rivers with perennial source of freshwater.

The northern most ranges of Himalaya are well known for their scenic beauty with lofty snow-covered mountain peaks garlanded by rivers and glaciers. Out of the mountain peaks of varying altitudes about a dozen of them rising up to 8000 meters above mean sea level (asl) or higher lead by the world's highest peak, the Mount Everest. The glaciers of Himalaya account for about 70% of the nonpolar-glaciated regions of the earth. The great south Asian rivers, the Indus with tributaries, and the Yamuna, Ganga, and Brahmaputra originate from the glaciers and mountain

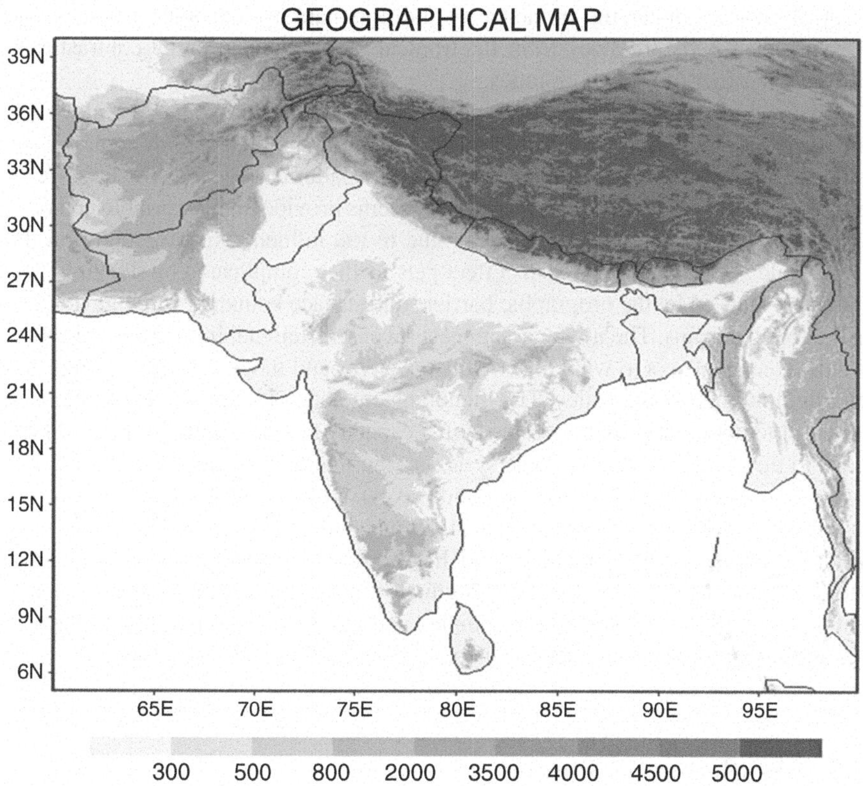

Fig. 2.1 Topographical map of the South Asian region depicting the Himalaya, the shades indicate altitude in meter

lakes of western and central Himalaya, while many tributaries of river Ganga arise from Nepal Himalaya. The river Brahmaputra traverses a long distance almost parallel to Himalayan ranges along its journey from central regions of Tibet to the Bay of Bengal through the mountains of north-east India.

Three major geological fold axes which constitute the Himalaya, Himadri (greater Himalaya), Himachal (lesser Himalaya), and Shiwalik (outer Himalaya), extend almost uninterrupted throughout its length. Geographically, the Himalaya generally described as an east-west mountain range has north-west to south-east orientation and the arc-like structure with enhanced southward slope to the east makes the eastern edges of Himalaya extending into eastern India and Myanmar. The western margins are located in Karakoram at about 37°N while the eastern ranges of Arunachal Pradesh are located around 27°N. Along its length the Himalaya can be divided into three sections: Western Himalaya (Hindukush Mountains, Jammu and Kashmir, Northern Punjab, and the Western Himachal Pradesh), Central Himalaya (Eastern Himachal Pradesh, Uttarakhand, and Western Nepal), Eastern Himalaya (Eastern Nepal, North-East India, Bhutan, and Myanmar). The enhanced southward slope to the east makes the eastern edges of Himalaya more close to the tropical weather under the influence of the dominating monsoons; whereas, the western most parts are away from the tropical influences and enjoy contrastingly different weather of middle latitudes.

The meteorology of high and extensive mountain regions like the Himalaya is a unique example of direct interaction of high altitude mountains with complex terrain and locally originating atmospheric weather patterns as well as the large-scale migratory weather systems. The mountain systems provide highly variable and relatively less predictable weather patterns due to the influence of orography on the large-scale atmospheric flow. This effect gets further complicated by dynamic perturbations caused by the orographic barriers and surface boundary forcings induced at the rough terrains. The mountains in general exert frictional influences on surface winds, create barriers to wind and weather systems and induce vertical ascents and gliding flows across the valley. The mountains interact with the weather systems at different temporal and spatial scales with a capacity to create substantial modifications to the characteristics of these systems. The interactions can be at the level of synoptic scale weather system (fronts and waves), mesoscale mountain valley circulation and gravity waves, forced vertical ascents due to orographic barriers to individual convective cells and the cloud bands. Some of the micrometeorological measurements made over a location (Nainital) in the central Himalayan region providing the information on how the complex and mountainous orography influences the surface winds are discussed in Chap. 3.

2.2 Weather and Climate

Large variation in topography, elevation, soil, rock structure, and vegetation cover result into contrasting climates within short distances in the mountain ranges along their extensive west-east spread. Broadly speaking, the Himalayan climate has prolonged winters and equally long summers, which include the monsoons in its southern periphery. Besides these local variations of weather, the Himalaya in general experiences a weather pattern dictated by the monsoon systems of Asia to its south and the weather patterns of the middle latitudes to its northern parts. The interface between the tropical and extratropical weather at higher altitudes create conditions highly conducive for high intensity extreme weather events. The transition seasons between winter to summer and summer to winter, akin to the spring and fall seasons of the extratropical regions, are not as distinct as experienced over the peninsular India, where they are often referred as pre- and post-monsoon seasons.

Proximity of Eastern Himalaya to Bay of Bengal accounts for the warm humid weather and an extended period of monsoon over the region compared to its western counterpart where generally dry weather prevails during the south-west monsoon, specifically to its western most periphery. The Himalaya functions as a great mountain wall dividing the climates of south and central Asian region. In winter, the great Himalayan ranges serve as effective barriers to the intensely cold continental air blowing southwards to the Indian subcontinent. These winds in the form of the north-east monsoon are blocked by the Himalaya from direct passage towards the south and are diverted further east providing some rain to the east coast of India in winter. This leaves rest of the Indian subcontinent practically free from these wind systems and winter rains. During northern hemispheric summer, the Himalaya forces the moisture bearing monsoon winds up the mountains to deposit most of their moisture over the southern slopes of the mountain ranges.

The Tibetan plateau at the top of the mountains acts as a high level heat source for large-scale monsoon circulation. The plateau with an elevation of about five kilometers is strongly heated by incoming solar radiation and is dominated by Tibetan high pressure belt. These conditions establish an anticyclonic circulation (clockwise) in the upper troposphere over a large area leading to easterlies all over the south Asian subcontinent at these levels. The Tibetan plateau with flat land and high elevation provides source of heat to the atmosphere with maximum in late spring and early summer. The intensity and interannual variability of its strength considerably influences monsoon activity over large parts of South Asia. The role of the Himalaya and Tibetan plateau system in generation and maintenance of South Asian summer monsoon is demonstrated in numerical simulation models by Godbole (1973) and Hahn and Manabe (1975). These simulations have shown that the monsoon circulation becomes practically absent when the mountain influence is removed from the model.

The snow-covered mountain ranges have high Albedo, an expression for whiteness or incident light. The albedo of a surface is defined as the ratio of reflected radiation from the surface to the radiation incident up on it, expressed as a dimensionless quantity. High albedo provides an important feedback to the climate system by providing a highly reflecting snow-covered surface for the incoming solar radiation which absorbs less energy and accelerates the growth of area under snow with extension of higher albedo area. It is generally inferred that the net albedo of the Earth is showing a decreasing trend as the snow and ice on the planet are melting as a result of the global warming.

As the white reflecting snow cover decreases in area, less energy is reflected into space and the warming of the Earth's surface gets enhanced. Over the oceanic regions, the disappearing ice while decreasing albedo triggers a positive feedback by exposing the oceanic surface to sunlight to warm the water. Keeping in mind the importance of the white snow cover in climate change and its numerous impacts on the Earth's surface, some scientists often suggest many impractical suggestions such as the spraying of black carbon shoot over the snow surface to alter the albedo.

Predicting the performance of monsoon rains over India on seasonal scale has been a challenging task for meteorologists, and it is interesting to note that the Himalaya has been the first to provide a predictor parameter to seasonal forecasts of monsoon rains. Henry Francis Blanford (British, 1834–1893 AD), the distinguished geologist and the Imperial Meteorological Reporter in India, appointed in the wake of the devastating cyclone of Bengal in the year 1864 AD was especially assigned the responsibility to develop the seasonal monsoon forecast. He put forth the hypothesis that varying extent and thickness of the Himalayan snow exercises a great and prolonged influence on the climate conditions and weather of the plains of north

India (Blanford 1884). The success of his tentative forecast during 1882–1885 AD for the monsoon season based on the Himalayan snow cover as leading predictor encouraged him to start the operational Long Range Forecast (LRF) of monsoon rainfall covering the then entire region of India and Burma in 1886 AD. The LRF efforts during the periods of severe and prolonged droughts over India in late nineteenth and early twentieth century using simple statistical correlations have unfolded many large-scale features of tropical circulations and the summer monsoon.

His successor Sir Gilbert Thomas Walker (British, 1868–1958 AD) following the severe monsoon failure of 1899 AD extensively studied the data over much of the tropical Oceans with focus on the Indian and Pacific region, the guiding principle being that the monsoon circulation may have large-scale global connections. These global influences subsequently became popular as the teleconnections. These efforts retained the influence of the Himalaya while adding a new parameter of oceanic origin to the LRF equations. This leads to the discovery of a see-saw oscillation of atmospheric pressure between the Indian and Pacific Ocean known as the Southern Oscillation which was later named after him as Walker Circulation (Walker 1918). His statistical analysis incorporating the data from global tropics and the Eurasian region established the foundation of the LRF of monsoon rains over India. Subsequent LRF for the monsoon season prepared by India Meteorological Department (IMD) used three predictors and the first among them remained the Himalayan snow cover for the period October to May. The practice continued till recently when multiparameter statistical models and dynamical general circulation models (GCMs) replaced these original prediction equations. However, the influence of Himalayan climate in the present day statistical and dynamical models of LRF continue to be represented as a manifestation of the tropical–extratropical interaction in some form or the other.

These early attempts during 1880–1920 AD in associating the snow cover over the greater Himalaya and deficient monsoon rainfall over India have distinctly emphasized the role of Himalaya in influencing the monsoon circulation. In the recent years, the availability of reliable and extensive satellite-based estimates of snow cover over Eurasia (Himalaya constituting the major part) have revived interest in the subject (Hahn and Shukla 1976; Dey and Bhanukumar 1983; Dickson 1984; Bhanukumar 1989; Saha et al. 2013) and broadly confirmed the existing concepts. It is highly significant to note that this concept has been under continuous validation for more than one and half a century and it is now accepted that the spatial extent of Himalayan snow cover during winter is an important slowly varying boundary condition for the subsequent development of monsoon circulation over the south Asia. Broadly speaking, an enhanced accumulation and extent of snow during the winter months is most likely to be followed by a weak monsoon (Parthasarathy and Yang 1995; Vernekar et al. 1995).

The parameters representing anomalous behavior of the sea surface temperatures of the southern Oceans more specifically represented by the indices of El-Nino and southern oscillation, a circulation anomaly over the Pacific and Indian Ocean region, now constitute the dominating climate feature on interannual scale. The climatologists commonly refer them as a combined term called ENSO. The ENSO therefore constitute an equally important predictor parameter having a complex but dominating relationship with the South Asian summer monsoon (Pant and Parthasarathy 1981). The other most important effect of the Himalaya on the summer monsoon over South Asia is the location of heat low (a low pressure area caused by the heated land and air over Sindh and adjoining areas prior to full establishment of the summer monsoon) and also the monsoon trough (lower atmospheric pressure belt with extensive convergence of moist winds at lower altitude levels) across north India. These low pressure centers along the convergence zone for moist air have their dynamical and thermodynamical origin. The location of trough axis along foothills of the Himalaya after its formation is influenced by the mountains and hill configuration and the orographic lifting helps in its buildup and sustenance. The mountains in general influence the atmospheric flow as a result of mechanical lifting and also through the heating due to condensation, which in turn is caused by air ascending over the mountain barriers.

The Himalayan mountain ranges are located in the subtropical high pressure belt where the seasonal north-south migration of pressure and wind systems generally alters the seasonal weather. In winter, the mid-latitude weather systems approaching from the Mediterranean region sweep over the mountain ranges and precipitation is abundant and mostly associated with the troughs and low pressure systems embedded in these circulations known as Western Disturbances (WDs) (Madhura et al. 2015). These winter circulations and disturbances bring in cold winds and precipitation in the form of rain and snow over north India which is the only source of snow to higher mountain ranges of the Himalaya. This influence is most prominent in Western Himalaya than the eastern one. The amount of snow depends on the altitude of the mountains and the intensity and frequency of the weather systems impinging upon them. The annual snow accumulation rate in the mountains of Western Himalaya west of 80°E and north of 34°N is reported to vary between 15 m and 1 m.

The snowfall in Himalaya generally begins in October and continues until April and May, with a maximum in January and February. Even during monsoon months, snowfall can take place over regions of higher elevations far up in the north. In the eastern Himalaya, these weather systems are generally clubbed with pre-monsoon thunderstorm activity due to the availability of higher amount of moist air originating from the Bay of Bengal and large-scale convective activity over the region. The snowfall here is mostly confined to the northern most high altitude mountain ranges with relatively smaller annual accumulation rate. In contrast to the winter season, the south-west monsoon season experiences a sweep of moist winds from the east forming a zone of convergence for the Bay of Bengal and the Arabian Sea branches of the monsoon currents over India. These are more vigorous over the CH and the eastern Himalaya. The rains weaken in strength as the monsoon currents progress towards the west losing most of its moisture before reaching its extreme western margins.

The rains during monsoon months (peak months being July and August) are rather heavy and mostly confined to the lower Himalayan ranges and foothills. Monsoon generally reaches the mountains of the WH in the first part of July though the pre-monsoon convection and orographic rains are also quite common during late summers. Because of direct access to the Bay of Bengal branch of south-west monsoon the eastern Himalaya experiences summer monsoon rains about a month earlier than the WH. The withdrawal of monsoon from the eastern Himalaya is generally a fortnight after it has started withdrawing from the WH in early September. During the peak monsoon months, the southern slopes of Himalayan mountains receive the maximum rain with eastern side exceeding in rain amounts compared to the regions far west. Thus, the difference in monsoon rainfall between the eastern and western portions of Himalaya is opposite to that of rainfall pattern experienced in winter.

There are large variations in monsoon rains on both sides on year-to-year basis, the interannual variability being much higher over the west as compared to the east. It is also observed that the rainfall patterns in the eastern and Western Himalaya are generally in opposite phase. Spatial distribution of rainfall over the country suggest that on many occasions a drought in west peninsular India is a years of excess rainfall and floods in the eastern parts and the vice versa. There are certain occasions when the monsoon reaches the western and central Himalaya much earlier than its normal date of arrival due to the faster than normal advance of the Arabian sea branch of the monsoon across the Indian peninsula. These monsoon currents get coupled with the WDs and local orographic/convective weather systems resulting into enhanced and prolonged rainfall activity. These are the occasions of heavy rain in Himachal Pradesh, Uttarakhand, and Nepal Himalaya resulting in flash floods in the rivers originating from the Himalaya and inundation of flat lands in adjoining plains.

During the situations of weak monsoon activity, when the monsoon trough is nearer to the foothills of Himalaya there is a temporary drought-like situation called break monsoon condition in central parts of Indian peninsula. During the early period of the break monsoon conditions, the lower Himalayan ranges generally get heavy rains resulting into flood in rivers due to channelization of moist monsoon current in a narrow band along the Himalayan foothills. The break monsoon condition may prevail for the duration of many days to few weeks and may also occur on more than one occasion during some years.

Himalaya displays large inhomogeneity in surface temperature distribution as well as in the vertical temperature profiles. In addition to normal vertical temperature lapse rates, latent heat in rain, snow and hail formation, and dissipation processes (evaporation, condensation, and sublimation) along with moist adiabatic processes (moist air being lifted up without losing its heat to the surrounding environment) induced during orographic ascent and descent contribute to these inhomogeneous conditions. At the ground level due to inhomogeneous topography and local weather systems, the temperature varies greatly from place to place and from hour to hour. Low density of air and intense direct solar radiation and variable cloud cover also play an important role in temperature variability across the Himalaya. Mountain areas are unique in the sense that during the daytime one can suffer from sunstroke and at night from acute frostbite on the same day (Mani 1981).

Within the large mountain ranges and valleys of the Himalaya, cool sunny weather turning within hours into unpredictable local shower or thunderstorm with heavy rain or hail is a common sight at many places. Some of the important features at a mountain location such as slope, aspect, local relief, lakes, waterfalls, thick vegetation cover, and such other features may give rise to many peculiar and highly localized weather phenomena with unique experiences. An east-facing mountain slope will have warm mornings with reduced fog and cooler afternoons, whereas the west-facing mountain slopes will experience the opposite effects. During summers, a river valley in between the mountains with lakes and water bodies may experience warm and humid weather with highly localized convective thunderstorms in the afternoon.

Pronounced temperature inversions (normal tendency of temperature decreasing upwards from the ground is reversed due to cold earth's surface with warmer layers of air above) are observed at some locations. Under these conditions, the air cooled by radiation at night flows down the slopes to the bottom of the valleys and settles in hollows or a thick layer of smoke and pollutants trap the heat of the surface air to create an inversion layer. In calm weather, undisturbed by winds, the stagnant air in the hollows is filled with fog when there is enough moisture with ground temperature inversion and creates a common site of valleys filled with white soft cotton-like appearance during the early morning hours. This fog dissipates within hours after the solar radiation evaporates it on clear days and the solar radiation reaching the surface lifts the inversion. Valley fog in mountain areas pose serious aviation hazard for landing and takeoff of airplanes due to drastically reduced horizontal and vertical visibility.

Temperature inversions due to nighttime cooling of the earth's surface also result into fog, mist, and dew over flat terrains. During morning hours with the rising sun, the radiation melts the dew and evaporates the mist and fog to carry the moisture upwards which forms the shallow layers of localized cumulus or stratocumulus clouds which disperses as the day passes. The diurnal range of temperature (difference between daytime maximum and nighttime minimum temperature) is generally higher at the bottom of the valley and lower on convex mountain tops than over the level terrain. The effect of changing climate on Himalayan environment and the contribution of changing Himalayan ecosystem to global climate are complex and least understood. An examination of the rainfall pattern in Himalaya, particularly the central Himalaya, where tropical–extratropical interactions of weather systems are more prominent suggest an increase in the instances of heavy rains and resulting flash floods in recent times usually associated with the WDs and enhanced local convection.

An example of recent happening is the devastating floods in river Mandakini originating from the glaciers at the top of Kedarnath valley in Uttarakhand. This devastating episode occurred due to the combination of meteorological and hydrological factors triggered by the cloudburst and heavy rains on the one side and the glacier lake overflow on the other side (Singh 2013; Dobhal et al. 2013). The monsoons had arrived in the region about a fortnight earlier in 2013 than its normal date of arrival, which was vigorous in nature and got coupled with the active western disturbances at those latitudes during the week of 15–19 June 2013 creating a meteorological condition conducive for very heavy precipitation and resultant floods which induced the glacier lake burst and the disaster which followed.

Due to paucity of surface and upper air observations on meteorological elements, glacier mass balance and snow extent and depth, the exact cause–effect relationships are difficult to establish for such exceptional events. The problem is more complex, as the cryospheric, atmospheric, and land surface processes are equally involved. Besides the permanent snow on the mountain peaks of the upper Himalayan ranges and many perennial glaciers, there is substantial accumulation of snow at lower altitudes and southern slopes of the mountains during winter months which melts in summer and provides large fresh source for groundwater recharge and streamflow. The amount of accumulated snow available as source of freshwater depends on the local climate, altitude of the mountain, and the intensity and frequency of weather systems.

2.3 Meteorological Observations Over Western Himalaya

Over WH, the network of conventional surface meteorological observations have very limited areal coverage and only few selected hill stations have long period of data. The stations with more than 100 years of continuous record on rainfall and temperature such as Srinagar (Jammu and Kashmir), Shimla (Himachal Pradesh), and Mukteshwar (Uttarakhand) provide highly significant meteorological records of climatic importance. Most of the Himalayan regions of great meteorological significance are located at higher altitudes with difficult terrain and rough weather conditions; hence, they are devoid of very useful long period meteorological records. Lack of data with adequate spatial and temporal coverage is now being recognized as a great deficiency and their augmentation are being taken up with priority using AWSs and satellite remote sensing. At present, it is difficult to document and display long period climatic features and their trends and variations over large spatial and temporal domains for these regions. The compilations of recent data over a period of few decades supplemented by satellite observations would be helpful in drawing definite conclusions regarding the magnitude of global warming over the region. Long period data on climatic elements such as precipitation (rain, snow, hail), pressure, temperature, wind, sunshine, incoming and outgoing radiation along with the level of greenhouse gases (GHGs) concentrations at the surface are required to be routinely measured and records to be safely archived.

In view of intense weather systems being directly related to vertical convection and atmospheric instabilities, vertical profiles of significant parameters are very much required for weather prediction and also to provide a three-dimensional climate picture. In addition to general climatic picture, the information and forecast on weather and climate parameters is essential for agriculture, water resource, tourism, natural disaster management and strategic civilian, and defense aviation services and operations. During the recent decades, multipurpose and multi-institutional efforts are underway to collect and assimilate the weather and climate data over the Himalaya using a combination of automatic and conventional surface observatories and remote sensing data platforms.

Most important among these are the programs initiated by the government departments particularly the India Meteorological Department, Indian Space Research Organization, Ministry of Defense, Ministry of water resources, and the Department of Science and Technology to bridge the data gap in mountain regions. IMD has a large network of various observational facilities across the peninsular India and adjoining coastal regions as explained in Chap. 1 but the Himalayan region still belongs to a very few of them. Meteorological satellites right through the INSAT-series, KALPANA, etc. from India to the most recent state-of-the-art versions around the world and other remote sensing techniques are being used to augment and improve the spatial coverage and quality of information. Remotely sensed data needs advanced retrieval techniques to account for complex orographic features and require validation with high quality ground observations. All diagnostic studies and input for prediction models are invariably using the grid point values of reanalysis data which have proved to be of immense value in analyzing the broad features of the weather systems.

There is an urgent need to enhance the facilities for vertical sounding of atmosphere using ground-based remote sensing tools such as RaDARs of appropriate wavelengths (Wind profilers, Doppler Weather RaDAR, cloud RaDAR) and LiDARs, supplemented with Acoustic Sounders and conventional Radiosonde (Radio signals from the balloon-borne sensors) observations. These observations are very important for the monitoring and forecast of extreme weather events. Special programs of observations in campaign mode need to be carried out in order to investigate and understand specific atmospheric processes. These are essential and of paramount importance to fill the data gaps and improve the quality, in order to support the meteorological operations and atmospheric science research over the region.

In view of cleaner and pristine air of the Himalayan region, observations of atmospheric GHGs, other trace gases and aerosols at high altitude mountain stations will also prove to be the useful benchmark for air quality monitoring. High precision ground observations and tropospheric vertical profiles integrated values for the entire column from the ground to the top of the atmosphere on all relevant parameters are required to be measured. These observations will have to be at many stations evenly located in the region to be recorded at least once a day or at mandatory time intervals as prescribed by World Meteorological Organization (WMO).

2.4 Challenging Scientific Issues

With better quality and coverage of data and improved understanding of weather and climate processes and availability of modern technology the prediction of natural disasters related to weather and climate extremes have improved to the level of reliability and general acceptance. A lot is yet to be done to improve weather and climate services in the mountains as large and diverse as Himalaya. Concerted efforts on part of the scientists, engineers, policy planners, and general public are required to identify the unique weather and climate-related issues of the region and find out appropriate scientific solutions. Following is a preliminary list of important features and challenging subjects of study relating to the Himalayan region, its weather, climate, and other related issues.

1. Study of convective and orographic storms, squall lines, cloudbursts, and other related meteorological phenomena which are associated with the local extreme

weather events often invigorated by the weather systems providing large-scale moisture convergence at ground level.

2. Natural disasters and environmental impacts of hailstorms, heavy snow, lightning, heavy rains, and extreme convective and orographic thunderstorms.
3. Flash floods, landslide, soil erosion, change in river course, loss of topsoil, and biodiversity.
4. Large-scale droughts and floods and their impact on water resources, agriculture, livelihood of the masses, transportation, hydropower generation, and communication. In this context, special efforts are required to model the specific cases of unusual behavior of south-west monsoons and WDs over the region.
5. Role of persistent dry and windy weather in the initiation and spread of forest fire with impact on biodiversity and forests, characteristics of atmospheric aerosols, air quality, and the local climate. Wind measurements and their vertical profiling may also be initiated at some sites with good potential for persistent strong winds for the purpose of wind power generation.
6. Mountain glaciers, snow-covered areas, and extent of seasonal snow and ice, and overall impact of cryospheric variability on local and global climate. Impact of climate change on glaciers and vice versa.
7. Forests, vegetation, human settlements, wildlife, and specifically the role of forests in reducing the global warming by acting as major pool of carbon sink.
8. Artificial channeling of river flow, construction of dams, reservoirs, mining, hydropower projects, and their environmental and hydrological impacts in the background of changing climate.
9. Mountain and valley fog, low clouds, turbulence, and their effect on different modes of aviation and ground transport.
10. Impact of mountain weather on agriculture, horticulture, livestock, tourism and socioeconomic impacts of impending changes under projected climate scenarios on an immediate and long-term perspective.
11. Role of Himalaya in the establishment and maintenance of large-scale circulation systems specifically the monsoons over South Asia.
12. Keeping in mind the long-term impacts of climate change detailed scientific investigations are necessary to map the sources of renewable energy, particularly the hydro, solar, wind, and geothermal energies over the mountain regions.

2.5 Trends and Variability in Climate over Western Himalaya

2.5.1 Altitude Dependency of Surface Climate Change Signal

While studying climate change at high altitudes, the issue of elevation dependency of surface climate change signal is very important primarily due to the following reasons (Giorgi et al. 1997). First, an enhancement in the changes in surface climate at higher elevations would imply greater impact on high altitude hydrology and ecosystem. Second, an amplified response at high elevation could be utilized as an early climate change detection tool. Third, the capacity of reproducing the elevation dependency of climate change signal could provide an important aspect of model verification. Climate change signals in particular the greater rates of warming at higher elevations compared to the low level ground areas have been reported from many parts of the world (Beniston and Rebetez 1996; Giorgi et al. 1997; Fyfe and Flato 1999; Beniston 2003: Pepin and Lundquist 2008; You et al. 2008, 2010; Kang et al. 2010) though there are few contradictory observations from some other parts of the world. Recent studies suggest a more sensitive response to global warming at higher elevations in China compared to other parts in the country (Titan et al. 2006; Yang et al. 2006; Kang et al. 2007).

Numerous arguments are put forth by the scientists to logically explain their finding on the elevation dependency of changes in surface climate. Beniston and Rebetez (1996) attribute it to the fact that high-elevation stations are more directly in contact with the free troposphere than the one at the low elevations; therefore, they are relatively less affected by ameliorating anthropogenic factors, such as urbanization and pollution. The snow-albedo feedback to climate system can also provide a strong elevation-dependent forcing. As snow is depleted at high altitudes under warming conditions, the surface albedo decreases, so that more solar radiation is absorbed at the surface and the surface warming is enhanced at the higher altitude thus providing a positive feedback (Meehl 1994). It is also important to note that the cooling history of the Himalayan ranges in longer time frame of geological periods are complex and largely a function of mountain uplift and erosion in addition to climate factors (Zeitler 1985). An overlap between geological and climatic influences may complicate the detection of global warming effect of very long timescales and its altitude dependency particularly in the systems in which the geomorphic features are continuing the process of formation, like in the Himalaya.

There are many ways of looking at the issue of climate change and its altitude dependency. In addition to changes in temperature, it is also possible to examine the influence on temperature extremes in relation to altitude as well as the slope and aspect of the mountain ranges. Analysis of data from some high-elevation stations on the Tibetan Plateau for the period 1961–2005 showed no significant correlations between elevation and trends and magnitudes of temperature extremes, except for coldest day temperature (You et al. 2008). In South America, lower elevations to the west of the Andes have experienced the greatest warming, while warming at higher elevations to the east is less marked (Vuille et al. 2003). An analysis of 1084 stations from the Global Historical Climate Network (GHCN) and data of the Climate Research Unit, University of East Anglia, Uttarakhand, by Pepin and Seidel, (Pepin and Seidel 2005) did not yield systematic relationships between temperature trends and the elevation.

Higher elevation sites over the south Asian region with monsoon dominated climate exert significant influence on climate by way of a permanent orographic barrier to the prevailing moist monsoon currents. This creates a steep precipitation gradient along the mountain slope as well as contrasting rainfall regimes along the windward and leeward sides of the mountain ranges as seen along the west coast of India during the peak south-west monsoon months. Higher monsoon rainfall at the mountain peaks may moderate the temperature profiles thus making it more complex problem to detect the effect of climate change on mountain temperatures. It is, therefore, concluded that the warming over these regions is more likely to be influenced by local factors and, hence, the changes in temperatures or their extremes may become less detectable and predictable (Revadekar et al. 2012).

An analysis of data over Nepal (Shrestha et al. 1999) suggests a relatively higher rate of warming over the country compared to the global mean which is attributed to contribution by higher rates of warming in high-elevation areas of the Nepal Himalayas and Middle mountain regions. Similar warming trends are also observed in the Tibetan plateau where the warming is more pronounced in higher altitude stations than in the lower ones (Liu et al. 2002). It is suggested that the reduction of snow and glacier cover in the Himalayas as a result of global warming may also be contributing to higher rates of warming observed in the higher elevation regions (Kadota and Ageta 1992; Yamada et al. 1992; Fujita et al. 1997; Jin et al. 2005). However, with limited available data it is not possible to arrive at a conclusion suggesting uniform elevation-dependent trend in mean temperatures or the extremes of temperature. Preliminary studies indicate that the Himalaya seems to be warming more than the global average rate and the temperature increases are greater during winter and the autumn seasons than during summers which are also larger at higher altitudes (Liu and Chen 2000; Shrestha et al. 1999).

The trends of change in rainfall over the Indian subcontinent have large spatial variability with patches of increasing and decreasing trends almost in balance resulting into a stable long-term monsoon rainfall regime. Model simulations of precipitation over the Western Himalaya in general indicate small but statistically insignificant increasing trend which is also detectable in the data for few stations with long series of available data over the region (Kumar et al. 2006; Agarwal 2009). There are trend analyses for the mean annual temperatures of few stations over the Himalaya with limited continuous data period. For example, Kumar et al. (2008) infer that the mean annual temperature in the Alaknanda valley (Western Himalaya) has increased by 0.15 °C between the years 1960 and 2000 and Sinha (2007) found that the average temperature of Kashmir valley has gone up by 1.45 °C over the last two decades.

Aerosol Optical Depth (AOD), a measure of the opaqueness of the atmosphere in a vertical column from the earth's surface with reference to specific part of the solar spectrum due to presence of aerosols, is taken as an index of change in regional climate. The AOD measurements are being done in situ at a few selected stations in many mountain regions with detection of climate change signals in mind. Recently, scientists have been using many satellite-based instruments to quantify the aerosol generated radiative forcing to the climate system. In Himachal Pradesh, at Kullu, a valley station the AOD obtained through Multi-wavelength Radiometer (MWR) has shown highest ever AOD at 500 nm wavelength as in May 2009 which was 104% more than mean AOD value from April 2006 to December 2009 (Kuniyal et al. 2009).

The value of AOD was found to be much smaller at Nainital, a hill station in Uttarakhand at higher altitude (Pant et al. 2006) thus suggesting a sharp regional variability in AOD and also in an important component of the radiative forcing to the climate. Temperature rise due to radiative forcing from aerosols in the atmosphere based on per unit AOD increase at Mohal-Kullu (Himachal Pradesh) was calculated as high as 0.95°K/day during summer (April-July) and as low as 0.51°K/day during winter season (December, January-March) (Guleria et al. 2011). There are meteorological observations over high altitude stations and glaciers in many part of the WH for selected seasons over a part of the year with discontinuities in the data due to many dates with missing observations. In view of their limited spatial coverage and discontinuities in time series, this information is of little climatic significance.

2.5.2 Temperature Changes Over Western Himalaya Since 1901

The region of Western Himalaya for the analysis carried out in the present study is a latitude-longitude domain. In view of the epochal behavior of the monsoon rainfall over India and also considering a period of 30 years as a representative time period in climatology, a detailed analysis of temperature over the Western Himalaya is done for 30 year subperiods using a homogeneous data for the period 1901–2007. The time series for maximum temperature during the months of January and February over the period from 1901 to 2007, along with the 31 sliding trend over the Western Himalayan region is given in Fig. 2.2. The monthly maximum and minimum temperature are analyzed for trends and variability for the following subperiods: (1) 1901–1930, (2) 1931–1960, (3) 1961–1990, and (4) 1991–2007 (Figs. 2.3, 2.4, and 2.5). For the above-mentioned periods, the annual cycles in the maximum and minimum temperatures are examined and compared to detect the signal of global warming (Figs. 2.6, 2.7, and 2.8). Basic characteristics of the annual cycles, the persistent increase from January to June and then decrease till the end of the year remains practically the same for all epochs. However, the most recent epoch of 27 years (1991–2007) shows substantial increase in both maximum and minimum temperatures.

In general, it can be inferred that the warming have been occurring prominently during the winter season and changes are smallest during the summer monsoon season (JJAS). The warming trend shows a clear increase for each epoch starting from the beginning of the twentieth century. Epochal trend analysis for monthly maximum temperature clearly shows a negative trend in maximum temperature during period 1931–1960 indicating cooling during the epoch (Fig. 2.2). The epochal trends in minimum temperature, on the other hand displays a negative trend during the period 1901–1930 (Fig. 2.3). The time series of both the maximum and the minimum temperatures show a positive trend during the recent period throughout the year from January to December. Higher trends are seen in all winter months compared to the rest of the months in a calendar year.

A characteristic feature distinctly seen in all epochs is the smallest value of trend as well as variability in maximum and minimum temperatures during the monsoon season. Though the most active period of monsoons in the Western Himalaya is during the months of July and August, the influence of local moisture with cloudy

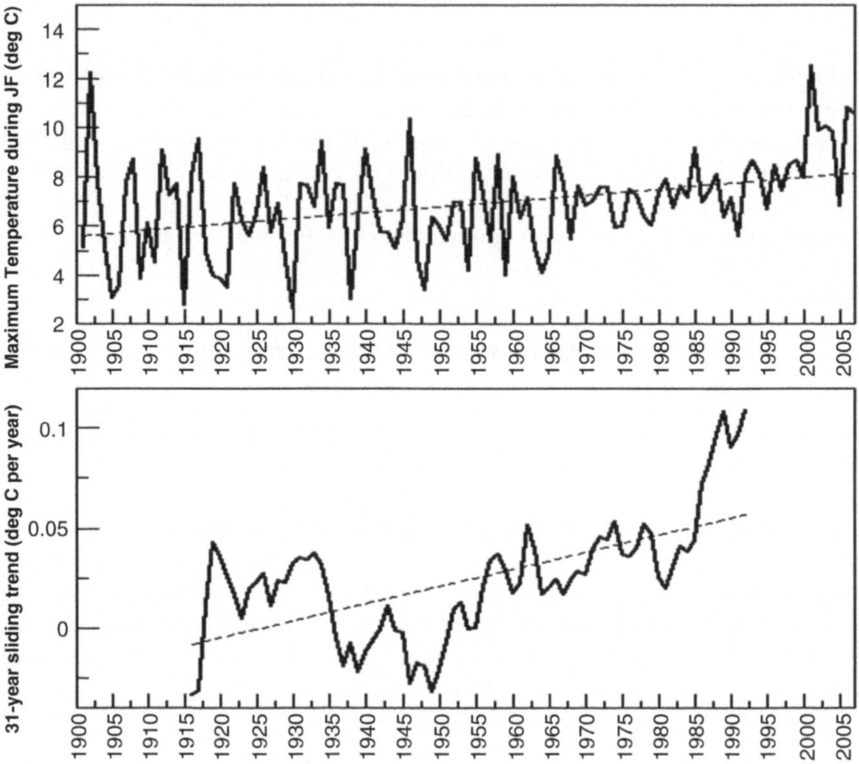

Fig. 2.2 Maximum temperature during January-February (*top*) and 31-year sliding trend (*bottom*) over Western Himalayan region, for the period 1901–2007

nights and strong summer convection coupled with orography perhaps moderates the rising summer temperatures throughout the monsoon season. During the summer months as the surface temperatures continue to increase, there is an increase in the intensity and frequency of organized convective activity particularly over Shiwalik and the southern slopes of the middle mountain ranges. The indications of the stabilizing effect of pre-monsoon convective activity and heavy monsoon rains on the hotter parts of the Western Himalaya are reflected in the moderating influence on monsoon surface temperatures.

Time series of seasonal temperatures constructed for Western Himalaya for the period 1901–2007 shows an increasing trend during all seasons, namely, January-February (JF), March-April-May (MAM), June-July-August-September (JJAS), and October-November-December (OND) for maximum temperature (Fig. 2.9) and for minimum temperature (Fig. 2.10). A summary of the results of the above analysis are presented in Table 2.1 for comparison at a glance. The details presented in

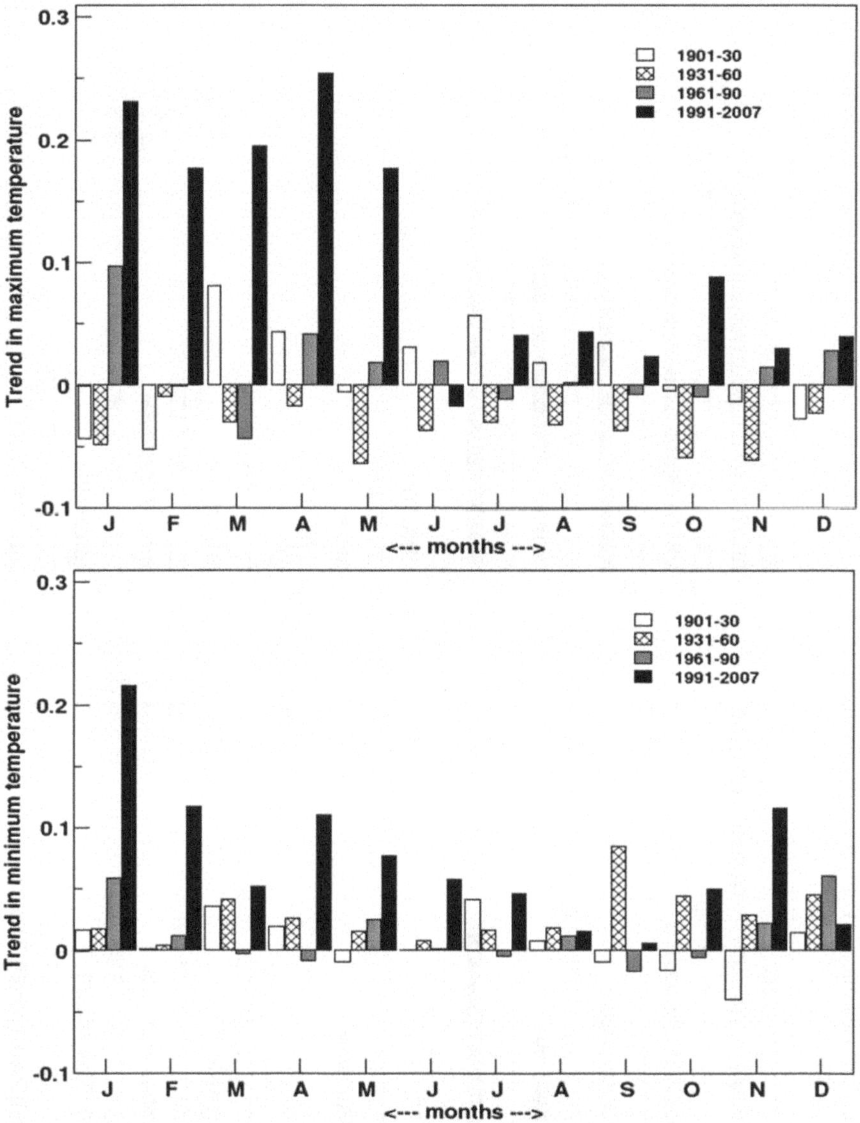

Fig. 2.3 Monthly trend values of mean maximum (*upper panel*) and minimum (*lower panel*) temperature over Western Himalaya for successive 30 year periods since 1901 (recent period ends at 2007)

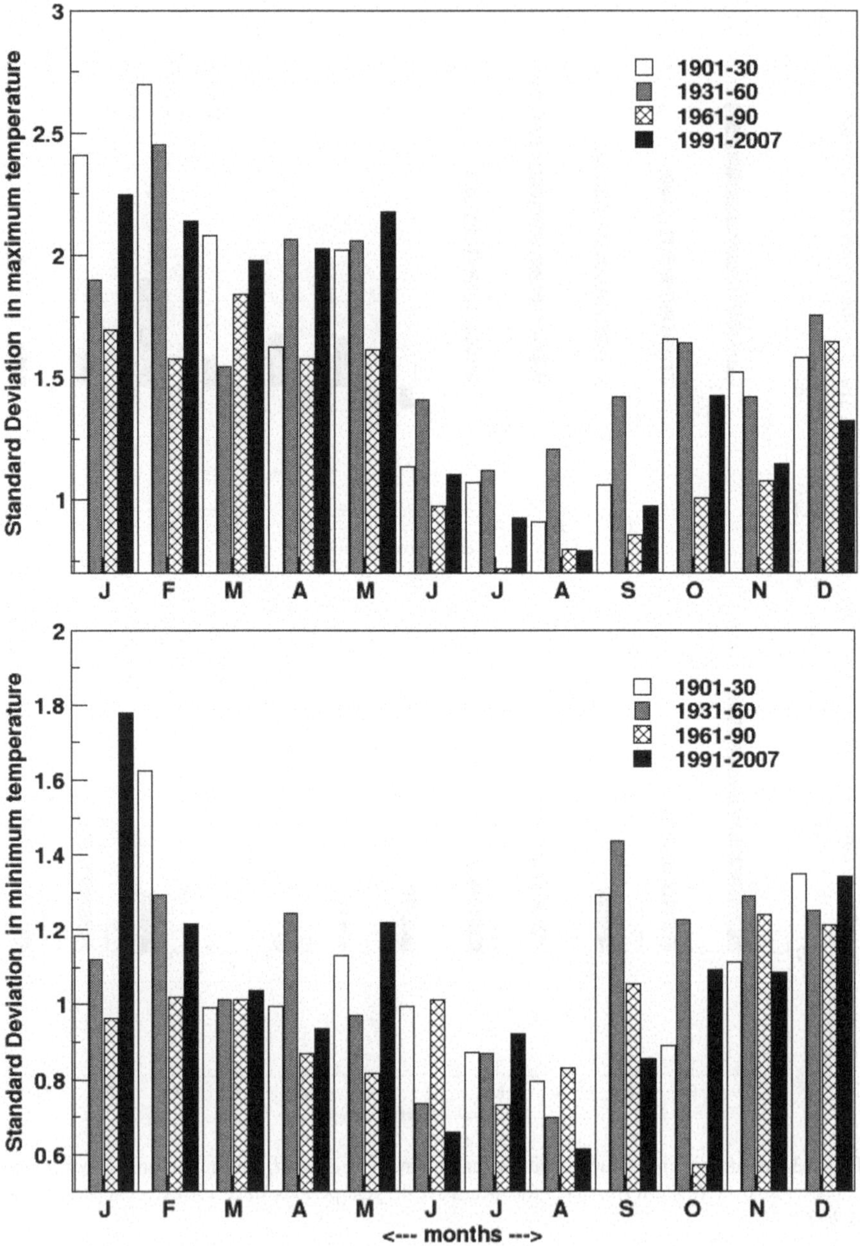

Fig. 2.4 Monthly values of the standard deviation of maximum (*upper panel*) and minimum (*lower panel*) temperature over Western Himalaya for successive 30 year periods since 1901 (recent period ends at 2007)

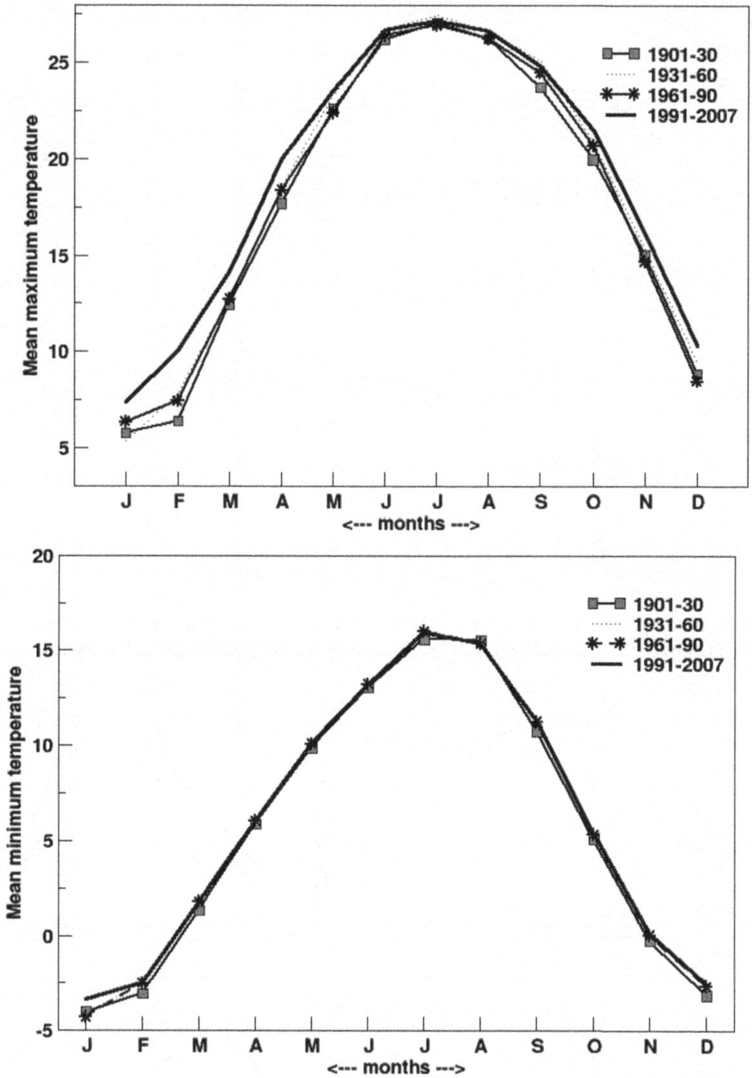

Fig. 2.5 Successive increase in monthly mean maximum temperature for 30 year periods 1991–2007

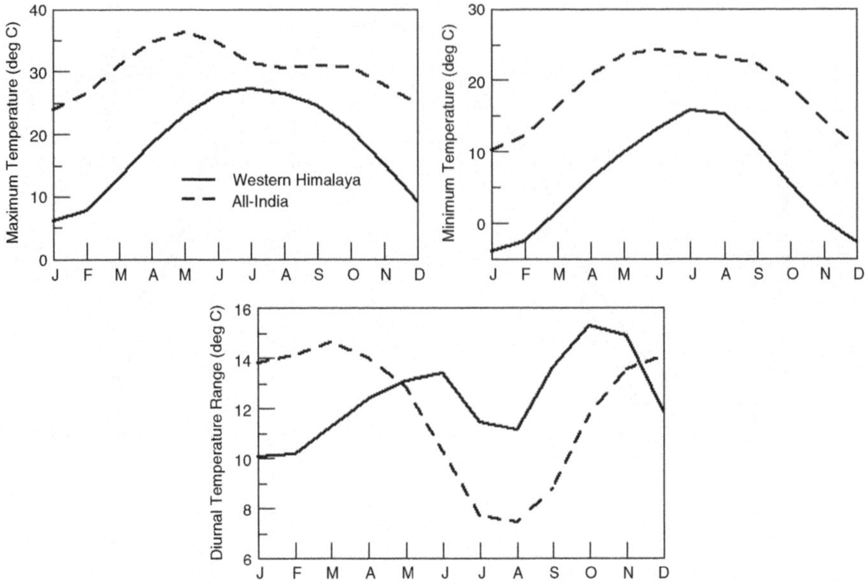

Fig. 2.6 Annual cycle of Maximum temperature, Minimum temperature, and Diurnal temperature range (max-min) for Western Himalaya (*solid*) in comparison to All-India value (*dash*)

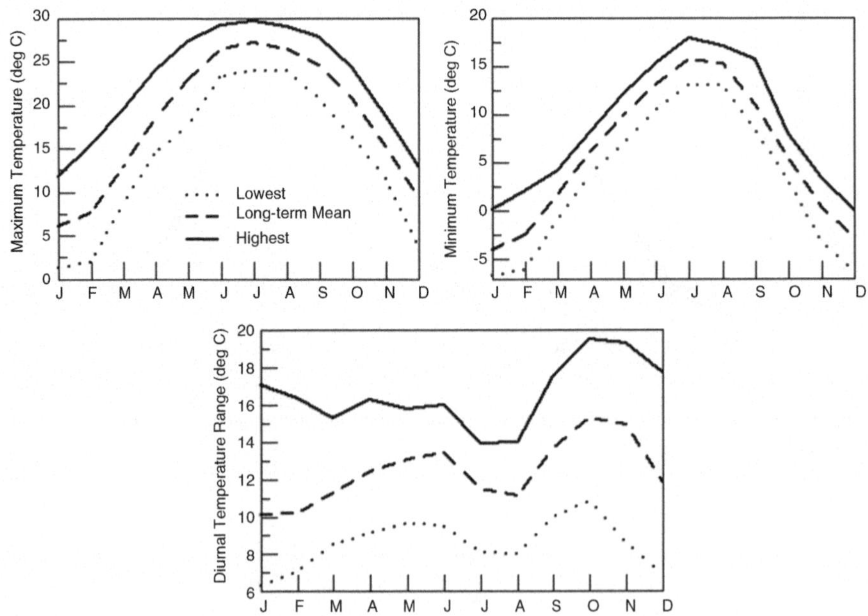

Fig. 2.7 Annual cycle of extremes in maximum, minimum temperatures, and temperature range for Western Himalaya in comparison to All- India values

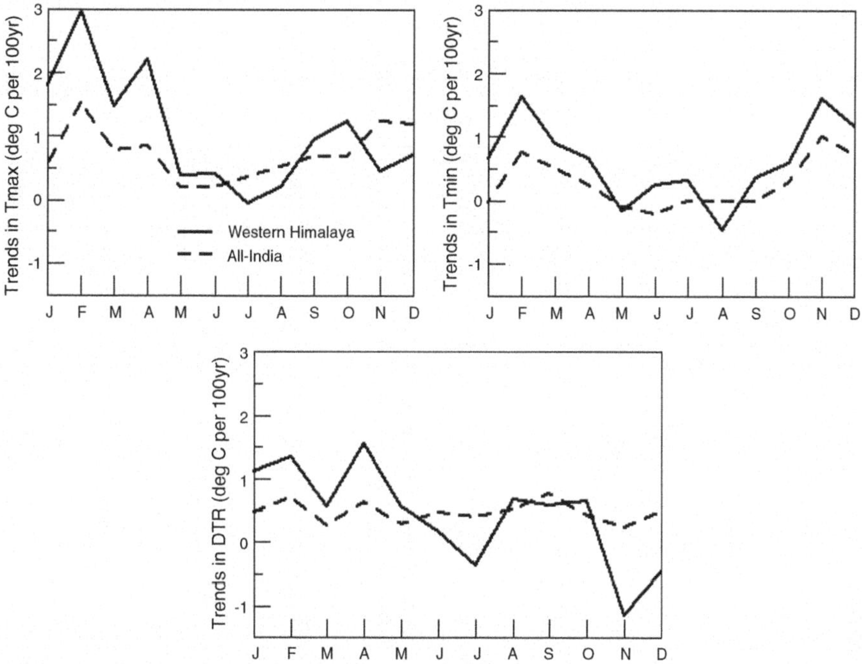

Fig. 2.8 As Tmax over Western Himalaya is showing higher trends during winter, JF time series is plotted (*top*) and 31-year sliding trends are computed (*bottom*). 31-year sliding trend show accelerated warming during recent period. Further analysis is therefore is done to find possible causes behind this warming over Western Himalaya

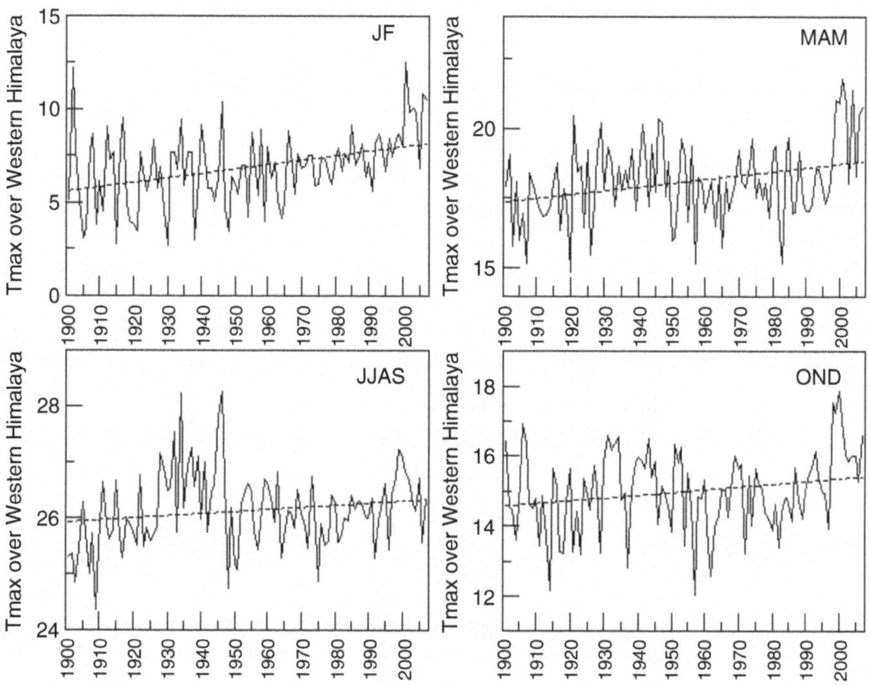

Fig. 2.9 Seasonal trends in maximum temperature over Western Himalaya

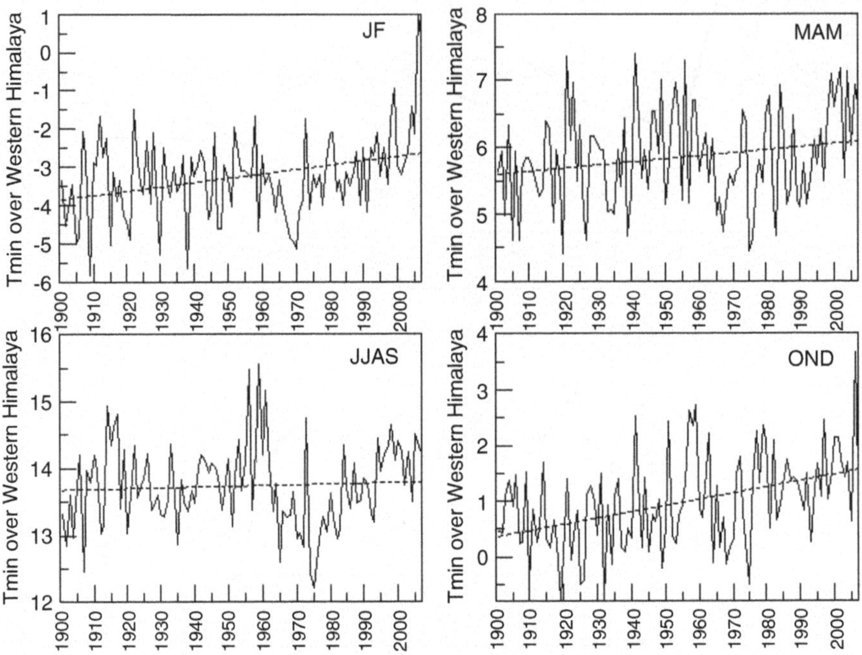

Fig. 2.10 Seasonal trends in minimum temperature in Western Himalaya

Table 2.1 Temperature statistics for Western Himalaya for the period 1901–2007

T_{max}	JF	MAM	JJAS	OND
Mean	6.87	18.09	26.23	14.99
Std. Dev.	1.97	1.43	0.69	1.14
Trend per decade	0.24[a]	0.14[a]	0.04	0.08[a]
T_{min}	**JF**	**MAM**	**JJAS**	**OND**
Mean	−3.26	5.84	13.74	0.97
Std. Dev.	1.12	0.74	0.63	0.84
Trend per decade	0.11[a]	0.05[a]	0.01	0.11[a]

[a]Indicates the statistically significant values at 90% level

the table and the analysis carried out for the monthly trends for maximum and minimum temperatures is based on the average taken over Western Himalayan region.

References

Agarwal PK (2009) Global climate change and Indian agriculture: case studies from ICAR Network Project. Indian Council of Agricultural Research, New Delhi, p 148

Beniston M (2003) Climatic change in mountain regions: a review of possible impacts. Clim Chang 59:5–31
Beniston M, Rebetez M (1996) Regional behavior of minimum temperatures in Switzerland for the period 1979-1993. Theor Appl Climatol 53:231–243
Bhanukumar OSRU (1989) Eurasian snow cover and seasonal forecast of Indian summer monsoon rainfall. Hydrol Sci J 33(5):511–525
Blanford HH (1884) On the connection of Himalayan snowfall and seasons of drought in India. Proc R Soc Lond 37:3–22
Dey B, Bhanukumar OSRU (1983) Himalayan winter snow cover area and summer monsoon rainfall over India. J Geophys Res 88:5471–5474
Dickson RR (1984) Eurasian snowcover vs Indian Monsoon Rainfall—an extension of Hahn-Shukla results. J Climatol Appl Meteorol 23:171–173
Dobhal DP, Gupta AK, Mehta M, Khandelwal DD (2013) Kedarnath disaster: facts and plausible causes. Curr Sci 105(2):171–174
Fujita K, Nakawo M, Fujii Y, Paudyal P (1997) Changes in glacier in hidden valley MukutHimal, Nepal Himalaya from 1974-1994. J Glaciol 43:583–588
Fyfe JC, Flato GM (1999) Enhanced climate change and its detection over the Rocky Mountains. J Clim 12:230–243
Giorgi F, James WH, Marinucci MR, Beniston M (1997) Elevation dependency of the surface climate change signal: a model study. J Clim 10:288–296
Godbole RV (1973) Numerical simulation of the Indian summer monsoon. Ind J Meteor Geophys 24:1–14
Guleria RP, Kuniyal JC, Rawat PS, Sharma NL, Thakur HK, Dhyani PP, Singh M (2011) The assessment of aerosol optical properties over Mohal in the northwestern Indian Himalaya using satellite and ground based measurements and an influence of aerosol transport on aerosol radiative forcing. Meteorol Atmos Phys 113. doi:10.1007/s00703-011-0149-5
Hahn DG, Manabe S (1975) The role of mountains in the south Asian monsoon circulation. J Atmos Sci 32:1515–1541
Hahn DG, Shukla J (1976) An apparent relationship between Eurasian snow cover and Indian monsoon rainfall. J Atmos Sci 33:2461–2462
Jin R, Li X, Che T, Wu LZ, Mool P (2005) Glacier area changes in Pumqu River Basi, Tibetan Plateau, between the 1970s and 2001. J Glaciol 51(175):607–610
Kadota T, Ageta Y (1992) On the relation between climate and retreat of Glacier AX010 in the Nepal Himalaya from 1978 to 1989. Bull Glacier Res 10:1–10
Kang S, Zhang Y, Quin D, Ren J, Zhanag Q, Grigholm B, Mayewski P (2007) Recent temperature increase recorded in an ice core in the source region of Yangtze River. Chin Sci Bull 52:825–831
Kang S, Xu Y, You Q, Flugel WA, Pepin N, Yao T (2010) Review of climate and cryospheric change in the Tibetan Plateau. Environ Res Lett 5. doi:10.1088/1748-9326/5/1/015101
Kumar KR, Sahai AK, Kumar KK, Patwardhan SK, Mishra PK, Revadekar JV, Kamala K, Pant GB (2006) High-resolution climate change scenarios for India for the 21st century. Curr Sci 90(3):334–345
Kumar K, Joshi S, Joshi V (2008) Climate variability, vulnerability, and coping mechanism in Alaknanda catchment, Central Himalaya, India. Ambio 37:286–291
Kuniyal JC, Thakur A, Thakur HK, Sharma S, Pant P, Rawat PS, Moorthy KK (2009) Aerosol optical depths at Mohal-Kullu in the northwestern Indian Himalayan high altitude station during ICARB. J Earth Syst Sci 118:41–48
Liu X, Chen B (2000) Climate warming in Tibetan Plateau during recent decades. Int J Climatol 20(4):1729–1742
Liu S, Shen Y, Sun W, Li G (2002) Glacier variation from maximum of the Little Ice Age in the Western Qilian Mountains, Northwest China. J Glaciol Geocryol 24(3):227–233

Madhura RK, Krishnan R, Revadekar JV, Mujumdar M, Goswami BN (2015) Changes in western disturbances over the western Himalayas in a warming environment. Clim Dyn 44:1157–1168. doi:10.1007/s00382-014-2166-9

Mani A (1981) The climate of the Himalaya. In: Lal JS, Moddie AD (eds) The Himalaya: aspects of changes. Oxford University Press, New Delhi, pp 3–15

Meehl GA (1994) Coupled ocean-atmosphere-land processes and south Asian monsoon variability. Science 256:263–267

Pant GB, Parthasarathy B (1981) Some aspects of an association between the southern association and Indian summer monsoon. Arch Meteorol Geophys Bioklimatol B 29:245–251

Pant P, Hegde P, Dumka UC, Sagar R, Satheesh SK, Moorthy KK, Saha A, Srivastava MK (2006) Aerosolcharacteristics at a high-altitude location in central Himalayas: optical properties and radiative forcing. J Geophys Res 111:D17206. doi:10.1029/2005JD006768

Parthasarathy B, Yang S (1995) Relationships between regional Indian summer monsoon rainfall and Eurasion snow cover. Adv Atmos Sci 12:143–150

Pepin NC, Lundquist JD (2008) Temperature trends at high elevations: patterns across the globe. Geophys Res Lett 35:L14701. doi:10.1029/2008GL034026

Pepin NC, Seidel DJ (2005) A global comparison of surface and free air temperatures at high elevations. J Geophys Res 110:D03104

Revadekar JV, Kothawale DR, Patwardhan SK, Pant GB, Rupakumar K (2012) About the observed and future changes in temperature extremes over India. Nat Hazards 60(3):1133–1155

Saha SK, Pokhrel S, Chaudhari HS (2013) Influence of Eurasian snow on Indian summer monsoon in NCEP CF Sv_2 free run. Clim Dyn 41:1801–1815

Shrestha AB, Wake CP, Mayawasky PA, Dibb JE (1999) Maximum temperature trends in the Himalaya and its vicinity: an analysis based on the temperature records from Nepal for the period 1971-94. J Clim 12(9):2775–2786

Singh DS (2013) Causes of Kedarnath tragedy and human responsibilities. J Geol Soc India 82:303–304

Sinha S (2007) Impact of climate change in the highland agroecological region of India. Sahara Time Magazine. http://www.saharatime.com/Newsdetail.aspx?newsid=2659

Titan L, Yao T, Li Z, MacClune K, Wu G, Xu B, Li Y, Lu A, Shen Y (2006) Recent rapid warming trend revealed from the isotopic record in Muztagata ice core, eastern Pamirs. J Geophys Res 111:D13103. doi:10.1029/2005JD006249

Vernekar AD, Zhou J, Shukla J (1995) The effect of Eurasian snow cover on the Indian monsoon. J Clim 8:248–266

Vuille M, Bradley RS, Werner M, Keimig F (2003) 20th century climate change in the tropical Andes: observations and model results. Clim Change 59(1):75–99

Walker GT (1918) Correlation in seasonal variations of weather. Q J R Meteorol Soc 44:223–224

Yamada T, Shiraiwa T, Iida H, Kadota T, Watanabe T, Rana B, Ageta Y, Fushimi H (1992) Fluctuations of the glaciers from the 1970s to 1989 in the Khumbu, Shorong and Langtang regions, Nepal, Himalayas. Bull Glacier Res 10:11–19

Yang X, Zhang Y, Zhang W, Yan Y, Wang Z, Ding M, Chu D (2006) Climate change in Mt. Qomolangma region since 1971. J Geogr Sci 16:326–336. doi:10.1007/s11442-006-0308-7

You Q, Kang S, Pepin N, Yan Y (2008) Relationship between trends in temperature extremes and elevation in the eastern and central Tibetan Plateau, 1961–2005. Geophys Res Lett 35:L04704. doi:10.1029/2007GL032669

You Q, Kang S, Pepin N, Fiugel WA, Yan Y, Behrawan H, Huang J (2010) Relationship between temperature trend magnitude, elevation and mean temperature in the Tibetan Plateau from homogenized surface stations and reanalysis data. Glob Planet Chang 71:124–133

Zeitler PK (1985) Cooling history of the NW Himalaya, Pakistan. Tectonics 4:127–151

Chapter 3
Weather Systems over Himalaya: Cloud and Precipitation Processes

3.1 Himalayan Weather Systems and Complex Topography

Climate over the Western Himalaya (WH) broadly comprises the winter and summer seasons with shorter transition periods in between. The winters in general are cold and mostly dry with occasional interruption by the weather system originating

G.B. Pant et al., *Climate Change in the Himalayas*,
DOI 10.1007/978-3-319-61654-4_3

over west Asia, the Mediterranean, and sometimes as far west as the Atlantic ocean. These weather systems on arrival; over WH are intercepted by mountain terrains giving rise to extensive cloudiness, rain, hail, and snow over the region generally followed by fog and severe cold wave (Pant and Rupa Kumar 1997).

The summers are very hot in plains of north India including the regions of lower mountain ranges and become mild as one moves towards the higher altitudes. During summer, the high temperatures and local heating along with orographic lifting often produces convective clouds of large vertical extent. There are several rivers and streams in the region, active even during peak summer months, which get recharged with melting glaciers and ground sources. These provide enough moisture to form local cloud systems resulting into high intensity rain events. The summer rains may start in early May and continue till September and sometimes even in October. The monsoon rains are rather heavy in lower mountain areas and valleys during July and August. The pre-monsoon and extended monsoon period provides a prolonged wet season over the region.

The Himalaya thus experiences a large variety of weather systems and associated complex atmospheric processes. Extensive mountain ranges and large display of topographic features play an important role in formation and development of orographic clouds. The microphysical processes inside cloud are complex and governed by the fundamental physical and dynamical law of nature. Many forms of water substances (hydrometeors) are produced under specific circumstances and conditions. All cloud formations during the passage of wind system over mountains may not produce some form of precipitation on the ground. The factors such as precipitable water content, buoyancy, height of cloud base from the ground, and depth of cloud system matter when the final precipitation product is realized. It is believed that the formation of convective clouds induced by orography is a major mechanism and can cause heavy precipitation over the Himalaya (Houze 1993).

Another weather event over this region is the fog, which is very common over the foothills of Himalayan mountain ranges and adjoining plains particularly during the winter season. Fog occurs when the air is cooled to near dew point and the water vapor in the air condenses on the aerosol particles present in the atmosphere. Two major types of fog, namely, the radiation fog and the advection fog are possible over the region depending upon the physical characteristics of the site and prevailing atmospheric conditions. Radiation fog occurs when the ground begins to lose heat in the night and the air above also cools near to the dew point temperature. In the presence of sufficient moisture, the water vapor condenses to form fog which stays for long time under calm or low wind conditions. The low level cloud with tiny moisture particles under weak wind conditions can be said as fog and the fog once lifted up from the ground may become low stratus clouds. The Himalayan region consists of a large network of rivers and an extensive biomass reserve with high soil moisture content, a necessary input to the formation of fog and hanging low clouds.

The second kind of fog is formed when the warm and moist air moves over to a cold surface and starts losing its heat to cool below the dew point which allows the moisture in the air to condense and to form tiny fog particles. This type of fog is called an advection fog, as it happens when the air is horizontally transported from one place to the other. The fog which occurs over a large area in the Himalayan foothills may persist for a much longer time affecting horizontal and vertical visibility, thus, becoming a potential hazard for aviation and road transport.

3.2 Microphysics of Clouds and Precipitation Process

While thinking of cloud microphysical processes in the Himalayan context, the first thing that comes to our mind is the snow. It is therefore essential to understand how it is formed and where it comes from. Snow is essentially produced in clouds with freezing temperatures around ice nuclei as a part of the microphysical cloud processes which also produce other precipitation forms. Before understanding the processes producing rain, snow, hail, and other precipitation forms, let us briefly explain the formation of clouds. Clouds in the atmosphere are formed when the air containing water vapor is cooled below its dew point and condenses on hygroscopic aerosol particles called cloud condensation nuclei (CCN) to produce small cloud droplets. Cloud droplets are generally in the size range of 10–100 μm (micrometer). The usual means by which air is cooled below its dew point occurs due to ascent and adiabatic expansion. There are a number of ways in which this ascent of air can happen. The most common types of ascent in the tropics is when due to local heating the air parcel can rise in what is called a conditionally unstable environment. Another type of ascent occurs when air gets lifted when passing over mountains. Then, there is the frontal lifting when two different air masses encounter which is important for mid and higher latitudes. Yet another type of lifting occurs due to horizontal convergence of air.

Rain forms when the cloud droplets are able to grow through collision and coalescence mechanism to produce millimeter size drops. A typical drizzle drop is of the size of 1 mm. The severity of rain events is classified by the amount of rainfall which takes place either over a 1 h period or a 24 h period. A heavy rainfall day as defined by India Meteorological Department is when the rainfall over a 24 h period is 64.5 mm or more (Guhathakurta et al. 2011). The precipitation formation for a convective and stratiform rainfall is different. The convective clouds are vertically much taller but their horizontal extend is less. The stratiform clouds are vertically shallow but horizontally cover a large area. In general, the rain from convective clouds can be very heavy and the duration is short as compared to rain from stratiform clouds which generally give persistent rain and drizzle.

When the temperature inside the cloud drops below freezing the cloud drops can freeze to produce ice. In the atmosphere, the freezing of the droplets without the aid of nuclei called as Ice Nuclei can happen only below −36 °C depending on the drop size. Ice crystals can directly form on ice nuclei by water vapor deposition. Most of the Ice Nuclei found in the atmosphere can activate at only temperatures below −5 °C. The probability of finding ice in the cloud between 0 and −5 °C is very less. Therefore, these unfrozen water drops which exists in the cloud at subzero temperatures are known as super cooled drops. As the number concentration of Ice Nuclei in the atmosphere is very less as compared to the CCN in the atmosphere, the super cooled drops in the clouds can be found till temperature isotherms of −40 °C. The Ice Nuclei concentration in the atmosphere is normally only one per liter of air and at lower temperatures can reach till ten per liter of air. These are far less as compared to the number concentration of CCN which goes to thousands per liter depending upon the pollution level in the atmosphere. Clouds in which both ice and super cooled water exist are known as mixed phase clouds and those clouds in which only ice exists, which happen when temperatures are lower than −40 °C, are called as glaciated clouds.

Once ice crystals have formed inside the clouds they can further grow by vapor deposition or by aggregation of ice crystals or accretion of super cooled drops. Ice crystals growing by vapor deposition in the clouds can take a variety of shapes (Fig. 3.1) which are controlled by the temperature regime and super saturation in the cloud. In Fig. 3.1 (a, b), the hexagonal plate on the right shows that first it had started growing as sector plate but then changed its shape to hexagonal. This is because of the crystal being exposed to different temperatures and water saturation during its growth process in the cloud. In Fig. 3.1 (c) and (e), dendrite growth is seen on the corners of plate crystal. This is because as the crystal is growing it will be exposed to different temperatures and supersaturation in the cloud, and therefore it can change its shape. Ice crystals growing by vapor deposition can grow to very big size at temperatures around −14 °C as at this temperature the maximum difference occurs between the saturated vapor pressure with respect to liquid water and with respect to ice. Ice crystals can serve as surfaces on which super cooled drops can collide and freeze and this mechanism of freezing is called as riming. Ice crystals on which drops have frozen and if the shape of the crystals is still visible, then it is called as rimed crystal. When the riming happens beyond a point where the shape of the ice crystal is no longer clear, then it is called a graupel. The graupel is usually of the size of a small pebble. If the graupel has been suspended in the cloud for a long time and has been growing by riming mechanism, then it can grow very big and it is called as hail.

Hail can form in vertically developing convective clouds with very high updraft velocities. A hail storm can be very damaging to both life and property as its fall speed will be very high as compared to the rain. In very vigorous storms, hail as large as 5 cm in diameter is possible and not uncommon. Hail can reach the ground even when the surface temperatures are 30 °C or more. Another mechanism of growth of ice crystals in the cloud is through aggregation. Ice crystals can collide with each other and form aggregates and resulting into snowflakes. The fall speed of snowflakes is very less and gentle when compared to rain or hail. For snow to reach the ground, the temperatures have to be subzero or near zero. The underlying microphysical mechanism of rain and snow formation in the clouds that form during the western disturbance is the same as that described above.

In the recent years, the Himalayan region has been experiencing heavy rainfall spells as well as cloudbursts. Cloudbursts are commonly understood as sudden breaking open of a dark thick cloud resulting into spontaneous pouring out of huge amount of water. In reality, it does not happen exactly that way. The cloudburst in simple terms is an event of short-term extreme precipitation over a small area that falls down as very heavy rain. Quantitatively cloudburst can be defined as an event in which 10 cm or more rainfall is recorded per hour over a place that is roughly

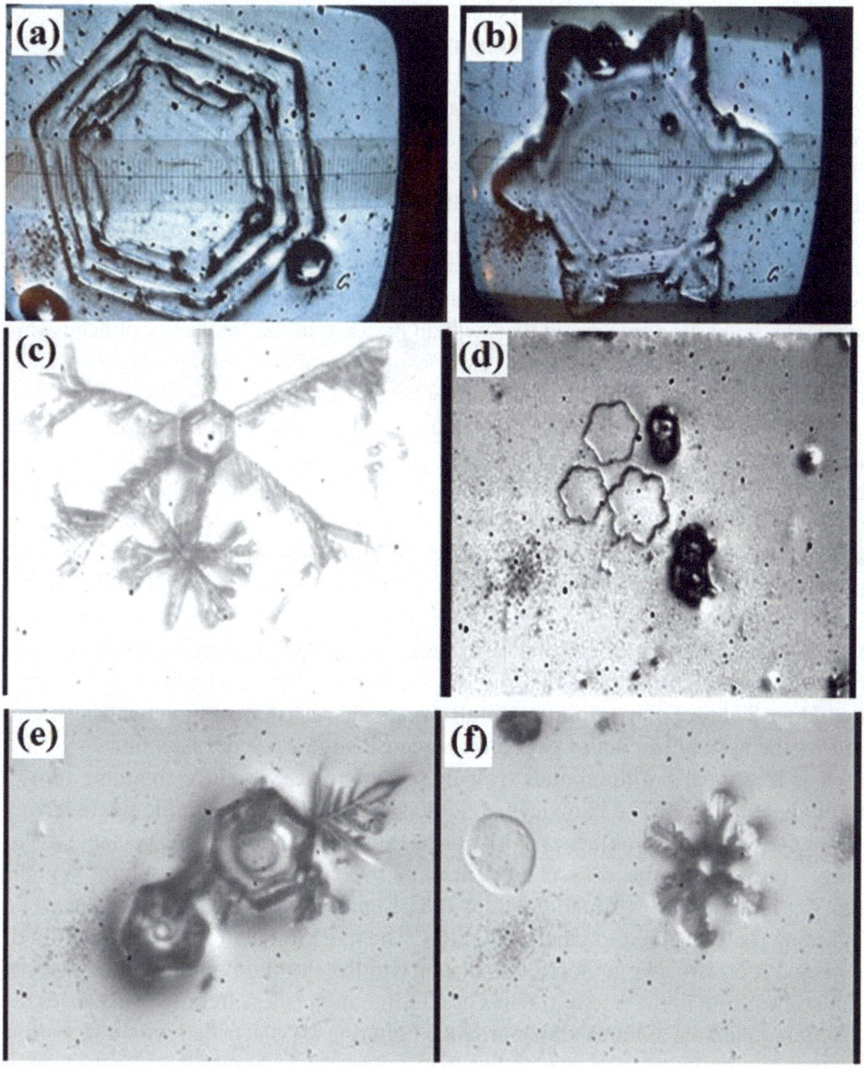

Fig. 3.1 Different shapes of ice crystal, (**a**) A hexagonal Plate, (**b**) Sector Plate, (**c**) Plate with dendrite growth growing from its corners, (**d**) Plate, (**e**) Dendrite growth on corner of hexagonal plate, and (**f**) Dendrites growing on all corners of hexagonal plate (**a**) and (**b**) 40× magnification, (**c**)—(**e**) 10× magnification

10 km × 10 km in area. The measure, therefore, is the rain rate over a unit area. The cloudbursts may happen anywhere, when the condition for localized heavy rainfall over a small area is satisfied; however, the probability of their occurrence in mountainous terrains is much higher. As an example, let us consider a situation when the fully saturated clouds are unable to produce rain where the warm current of air pushes the raindrops upwards and these drops are carried up with the current instead of dropping down. In this process, new raindrops are formed and existing drops

become bigger and heavier. After certain stage, the raindrops become too heavy for the clouds to hold and they start dropping down in heavy flashes.

In the mountainous terrain of the Himalaya, during pre-monsoon conditions, there exists a high probability of warm moist current rising vertically and on being propelled further up on encountering a mountain often results into cloudbursts. Being a highly localized event with probability to occur over an extensive area, it is difficult to pinpoint when and where will they occur. With advanced monitoring and prediction tools, advanced warning for heavy rainfall are being provided by the meteorologists. Another question about whether their frequency is increasing in recent years is much more difficult to answer since reliable data of the past is not available. It can be conjectured that with an increase in the frequency of heavy rain events in many parts of the world, as a result of global warming, the chances of increase in cloudbursts in the mountain terrains of Himalaya may also increase.

Cloudbursts can cause flash floods and in mountainous regions can cause heavy destruction. At the time of cloudburst rain rate can be in the range 200–1000 mm per hour (Joshi and Kumar 2006). In 2013, between 15 and 17 June the heavy rainfall occurred around the Kedarnath area of Uttarakhand state, which brought large amounts of debris from the glaciers and washed away the Kedarnath town. The 24 h precipitation recorded at the Chaurabari Glacier Camp of the Wadia Institute of Himalayan Geology was 325 mm in 24 h (Dubey et al. 2013) which can be classified as a cloudburst event. Rainfall in excess of 250 mm during the years 1871–2007 has been recorded mostly to the south of the Greater Himalayan ranges (Nandargi and Dhar 2011). Such extreme rainfall events can cause heavy landslides, which throws the normal life out of gear and can cause large-scale devastation.

It is observed that the clouds over these regions can produce lightning almost throughout the year. The Himalayan region is one of the regions in the world having the highest lightning frequency. Most of us think of lightning as a spectacular display of fireworks in the atmosphere. There are two broad categories of lightning, the first one is the cloud to ground lightning (CG-Lightning) and the other is the in-cloud lightning (or also cloud to cloud lightning). The most frequent is the in-cloud lightning and it covers a large horizontal distance in the sky illuminating the whole sky sometimes from horizon to horizon. The CG-lightning is less frequent but causes the maximum damage. Clouds which produce lightning are electrified with net positive charges accumulating at the top of the cloud and negative charges towards the center/ or base of the cloud. The accumulation of the charges results into development of very large electric field causing the electrical breakdown of the air, wherein the air becomes a conductor. Lightning in simple term is the electrical breakdown of air where a large current flows from one region to another. The credit for the earliest scientific observation on lightening in the sky is given to famous statesman and physicist Benjamin Franklin (American 1706–1790) and his kite experiment.

In a terrain like the Himalayas, the frequency of CG-lightning can be higher than that over plains, as the distance between the cloud base and the land surface is less over mountainous regions. The CG-lightning starts from the base of the cloud and comes down in steps towards the earth, each step of few hundred meters long with a pause of a few microseconds and is called the stepped leader. The stepped leader usually approaches the tallest object in the region. Objects like tall buildings, water tanks, trees, towers, electric poles, etc. are the most likely targets for lightning strikes. As the stepped leader approaches the striking point a return stroke goes to meet this stepped leader which transports a large amount of charge from the ground to the cloud. This stepped leader–return stroke combination constitutes a lightning discharge. The current flowing in the channel can be as high as 10,000–50,000 amperes. Such a large current flowing in a fraction of second in a channel whose width maybe only a few meters in diameter causes the temperature in the channel to increase. The channel temperature can go as high as 30,000°K. This sudden increase of temperature in the channel will cause pressure inside the channel to increase and therefore expand causing shock waves, which ultimately we hear as thunder. It is this large current and high temperature which can burn or crack the object on which lightning strikes.

The mechanism of how such high charges develop inside the cloud is still not fully understood. However, observations of lightning and cloud types shows that lightning occurs only in tall clouds which have a mixed type of precipitation. Mixed type of precipitation means those clouds in which water drops, ice crystals, graupel, hail, and snow are present. Lightning takes place in clouds which have developed very high electric field. The breakdown electric field which is required for lightning to take place is of the order of 300 kV/m. Saunders (2008) has given the requirement of thunderstorm charging as (a) time available for electric field generation is about 30 min, (b) charge generation per lightning flash is of the order of 20–30 C, (c) charge separation occurs between the 0 and −40 °C temperature levels in a region of 2 km inside the cloud, (d) the main negative charge center is between −5 and −25 °C temperature levels and the positive charge center a few kilometers above the negative charge center, (e) the electric field development is associated with the formation of graupel in the cloud.

Although there are several charging theories in the literature, most of the research has been done on the ice crystal–graupel collision charging mechanism which was initiated by Reynolds et al., (1957) and then followed up by Takashi (1978), Jayaratne et al. (1983), Pradeep Kumar, and Saunders (1989). Though lightning is one of the most powerful atmospheric hazards, deaths reported due to lightning from the Himalayan region is far less as compared to some other regions in the plains of India. The probable reason could be the population density less over the region. As stated earlier that mountain topography favors the lifting of moist air along the slopes and which can give rise to the formations of occasional clouds and thunderstorms. The influence of topography and the interaction of subtropical and mid-latitude weather systems over complex mountain terrain are described briefly in the subsequent paragraphs.

3.3 Influence of Mountain Orography on Surface Layer Winds in the Himalaya

Mountains are endowed with the capacity to develop their own characteristic local weather system through the influence it creates on the overlying atmosphere. These local weather systems are in a way representative of the geographical location, topography, and land cover characteristics. The competing forces of large-scale synoptic weather systems of distant origin and locally induced weather over mountains present a wide spectrum of atmospheric conditions. Mountain wind system is the main driver of local atmosphere with diurnal changes in their strength and direction. The mountain wind system can be in four forms, namely, slope winds, along valley winds, cross valley winds, and mountains-to-plains winds. These local wind regimes are largely influenced and modulated by the synoptic weather systems. The diurnal variation of surface layer characteristics in terms of certain relevant meteorological parameters observed near a mountain ridge during spring (March–May) of 2013 (Solanki et al. 2016) are briefly illustrated below as an example. The site for the data used in this chapter is the Manora Peak (29.4°N, 79.5°E, and 1926 m, amsl) belonging to the Aryabhatta Research Institute of Observational SciencES (ARIES), Nainital, south-central Himalaya.

During spring, this region generally witnesses fair-weather conditions and significant solar heating of the surface, providing favorable conditions for the systematic diurnal evolution of the atmospheric boundary layer. The three-dimensional wind components and virtual temperature observed with sonic anemometers (sampling at 25 Hz) mounted at 12-m and 27-m height on a meteorological tower have been used to study the influence of orography along with the solar radiation. It was found that, notwithstanding the prevalence of strong large-scale northwesterly winds, diurnal variation of the mountain circulation is clearly discernible with the strengthening of wind speed and a small but distinct change in wind direction during the afternoon period. Such an effect further modulates the surface layer water vapor content, which increases during the daytime and results in the development of boundary-layer clouds in the evening. An example of such an interaction observed through meteorological measurements taken over this mountain peak is depicted in Fig. 3.2 that indicates how the direction varies over a diurnal cycle as a result of the orographic influence and with the movement of sun around the valley.

The figures have been extracted from observation made in clear sky conditions, for the spring season and winter season of the year 2013–2014. It is clear from the figure that the wind speeds are rather strong (6 ms^{-1}) during spring season as compared to the weak (2–3 ms^{-1}) wind conditions in winter. It is well known that under weak synoptic conditions the diurnal variations in mountain wind system can be clearly discerned. One such locally observable system is the slope winds, consisting of upslope wind during daytime and downslope in nighttime. However, the time of transition between these two flows and their strength varies considerably from one mountain to other and the location at which observations are being made. In this case, the variations in wind direction and speed provide an insight into the wind system of Manora peak, when for the spring season wind direction changes to westerly flow in afternoon and returns back to northwesterly flow in nighttime. This, in fact, is the synoptic flow over the region which gets modulated by the upslope winds of mountain during daytime, also resulting in a minima in wind speed at 09.00 h. For the winter season when the synoptic flow is rather weak, a complete reversal in wind direction is observed in tandem with the change in the heating of the mountain slope during the course of the day. The northeasterly flow of nighttime is the result of downslope flows, whereas the transition from easterly to southerly and to westerly flow is the outcome of upslope flow.

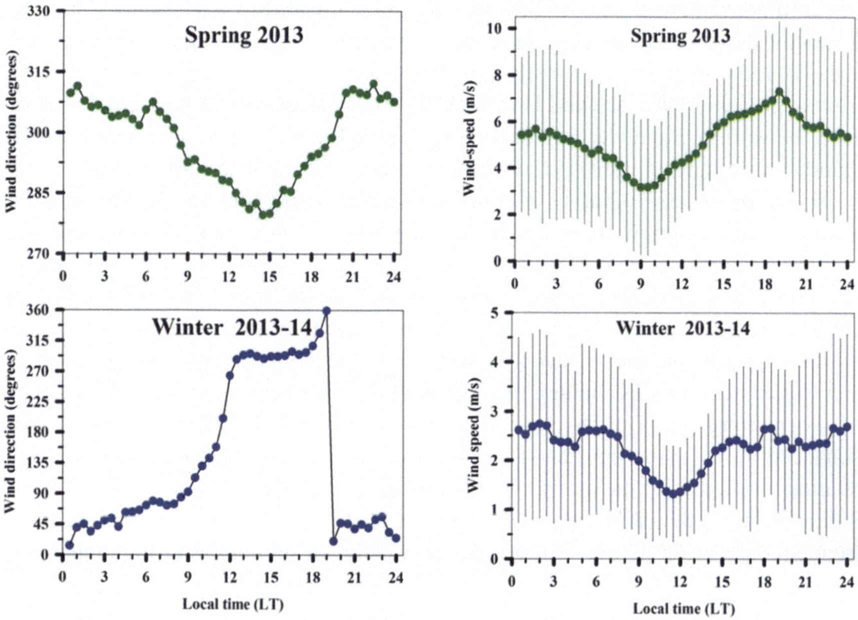

Fig. 3.2 Seasonal average wind speed and wind direction over diurnal cycle observed at Manora peak, Nainital during the year 2014–2015. Bars over wind speed panels indicate the standard deviation

3.4 Occasional Fronts Observed Over a Central Himalayan Site during Different Seasons

Atmospheric activity over Himalayan region influences the weather conditions of north Indian region all year round. The baroclinic fronts, a major attribute of mid-latitude weather visiting the Himalayan region manifest into variety of events, such as abrupt wind shifts, enhanced wind speeds, and turbulence. These fronts are formed as a result of atmospheric instability in which the displacements consist of warm air rising and moving poleward at some longitudes and cold air sinking and moving equatorward at other longitudes (Schneider 1996). The frontal zone is an interface of discontinuity in meteorological factors between two contrasting air masses with strong baroclinity (an atmospheric condition in which density depends on both the temperature and the pressure), higher absolute vorticity and vertical wind shear (Bond and Fleagle 1985; Schultz 2006). In cold fronts when two different air masses interact, moist air rises and form clouds, horizontal wind speed start accelerating with sudden change in direction. Under its influence, heavy thunderstorms are often witnessed with snow, rain, and wind gusts dropping temperature by as much as 10 °C.

Warm fronts on the other hands are at the leading edge of homogeneous warm air mass, located towards the equatorward edge of sharp thermal gradient, and lie within the broader troughs of low pressure as compared to cold fronts. This also forces the temperature differences across warm fronts to be broader in scale. Clouds ahead of the warm front are mostly stratiform, and rainfall gradually increases as the front approaches (Wallace and Hobbs 1977). Fog may occur in the wake of it, however, the warming and clearing of the fog is usually rapid after the frontal passage, thunderstorm may continue if the warm air mass is unstable (Schultz 2005).

The influence of orography in Himalayan weather and climate is observed in the form lee slopes, cyclogenesis, and local changes in the wind, temperature, and pressure. These are in agreement with growing evidence of the role of the earth's orography in weather and climate in the lower as well as the upper troposphere (Wilby 1995; Tong et al. 2009). The best example of tropical–extratropical interaction and most likely affecting the monsoon in northern India may be the cold wind advection and frontal activity approaching the central Himalayan region from the west during persistent long spells of Western Disturbances (WDs). The frontal activity with enhanced snow in Himalaya and the activity of monsoons over India may have some relation for predicting the monsoon on seasonal scale as discussed in earlier sections. Following sections are devoted to the description of some observational evidences of frontal passage over a high altitude central Himalayan station to illustrate the point.

3.4.1 Fronts Observed During Winter

Fronts approach north western Indian region at much lower latitudes during winter months (December, January, and February) in the form of transient weather disturbances. These disturbances largely derive their energy from the eastward moving synoptic waves and appear as the manifestations of the growth, decay, and propagation of instabilities in the atmosphere. During these episodes, the cold air mass suddenly replaces the warm air with rapid fall in temperature of the order of 5–10 . During the occurrence of these fronts, general warming of lower troposphere is followed by very cold winds in the rear characterized by sudden change in wind direction and speed. This causes extensive cloudiness, heavy precipitation, strong winds, and intense cold conditions. A classic example of changes in various meteorological parameters at Manora Peak on 23 February, 2013 at a height of 27 m from the ground level, is presented in Fig. 3.3.

In an observation plot of the day, there is a gap between peak activity hour starting around 09:30 hours (LT) to around 13:30 hours (LT) in which the instruments were shut down due to lightning and bad weather. Starting with the emergence of the activity around 08:30 hours (LT), the temperature suddenly dropped by about 6° and the wind direction abruptly turned to northeasterly from westerly. The horizontal wind speed received an accelerating push from a speed of 3 ms^{-1} at around 03:00 hours (LT) and reached a speed up to 20 ms^{-1} along with an increase in humidity which acquired 100% mark at the time of frontal onset. The vertical wind speed was recorded to be of the order of 4–5 ms^{-1} at the start of activity till the recording time. This front has caused heavy rains and thunderstorm with lightning and lasted for almost 6 h (09:30–1530 h, LT).

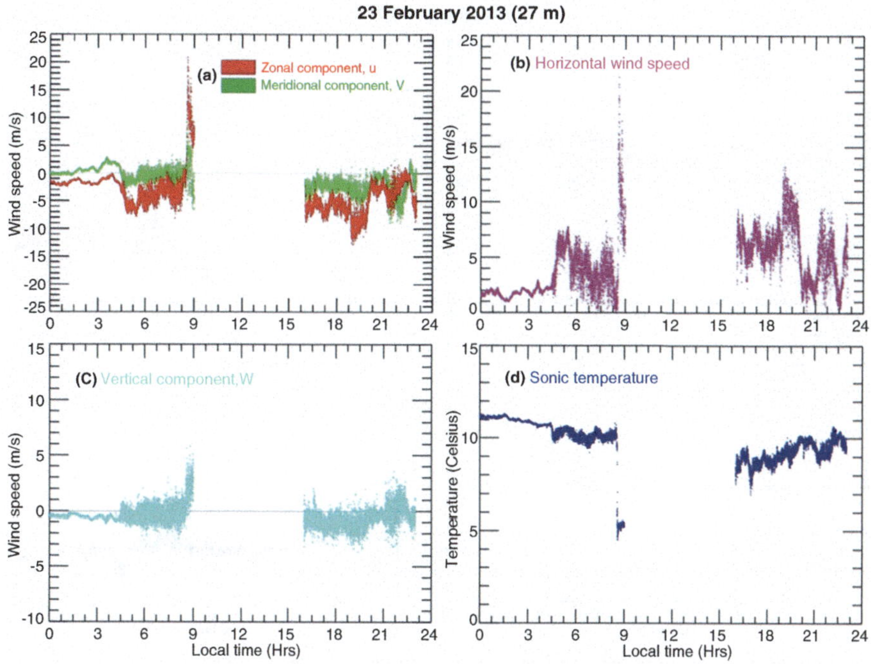

Fig. 3.3 High resolution measurements on diurnal variability of horizontal wind components: zonal and meridional (**a**), horizontal wind speed (**b**), vertical wind component (**c**), and sonic temperature (**d**) on 23 February 2013 (27 m)

3.4.2 Fronts Observed During Spring

At Manora peak, a maximum number of fronts were observed in 2014, with an activity in every month of this season (March, April, and May). First front was observed on 5 March in the evening hours. The activity started with the sudden increase in wind speed and a rapid drop in temperature by about 4° within 20 min. The winds accelerated to 15 ms^{-1} in horizontal and 4 ms^{-1} in the vertical. The weather parameters recorded during this frontal activity lasting for about two and a half hours is presented in Fig. 3.4.

The second front of the season was observed on 2 April during nighttime. This was an interesting event from the point of view of the building up of kinetic energy. The day started with high wind speeds in the morning starting around 05:00 hours (LT). There was a sudden acceleration of wind reaching up to 25 ms^{-1} by mid night. The wind and temperature parameters during the 24 h relating to this event which lasted almost 3 h are presented in Fig. 3.5.

In the month of May, weather is warmer with temperatures around 24°C over the station during daytime with dry conditions and clear skies. An evening hour frontal activity started at 21:00 hours (LT) on 8 May of the year with rapid drop in temperature by about 5° within an hour and wind speed gained strength. The highest wind

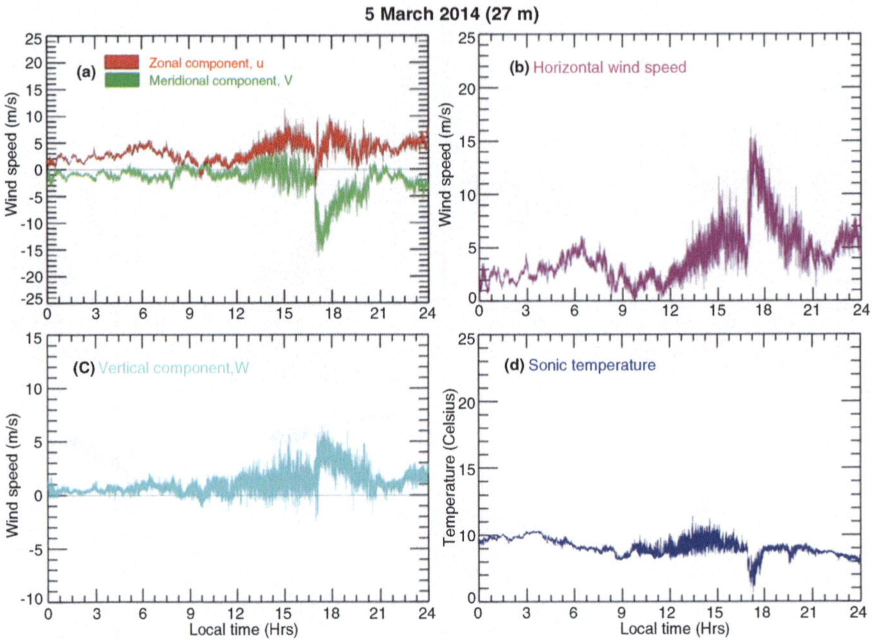

Fig. 3.4 Diurnal variability of zonal and meridional wind components (**a**) horizontal wind speed (**b**), vertical wind component (**c**), and sonic temperature (**d**) on 5 March 2014 (at 27 m level)

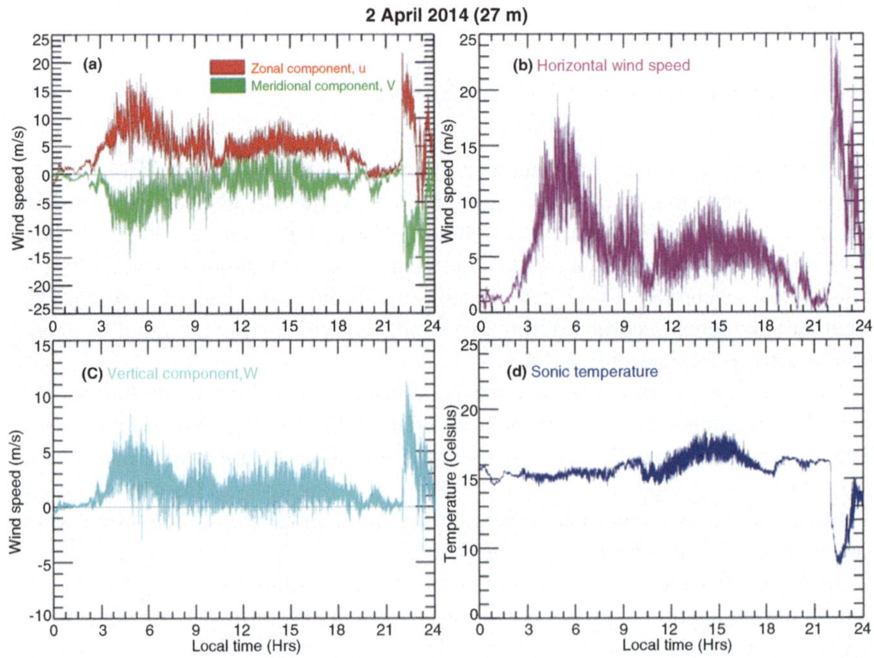

Fig. 3.5 same as Fig. 3.4 but for 2 April 2014 (at 27 m level)

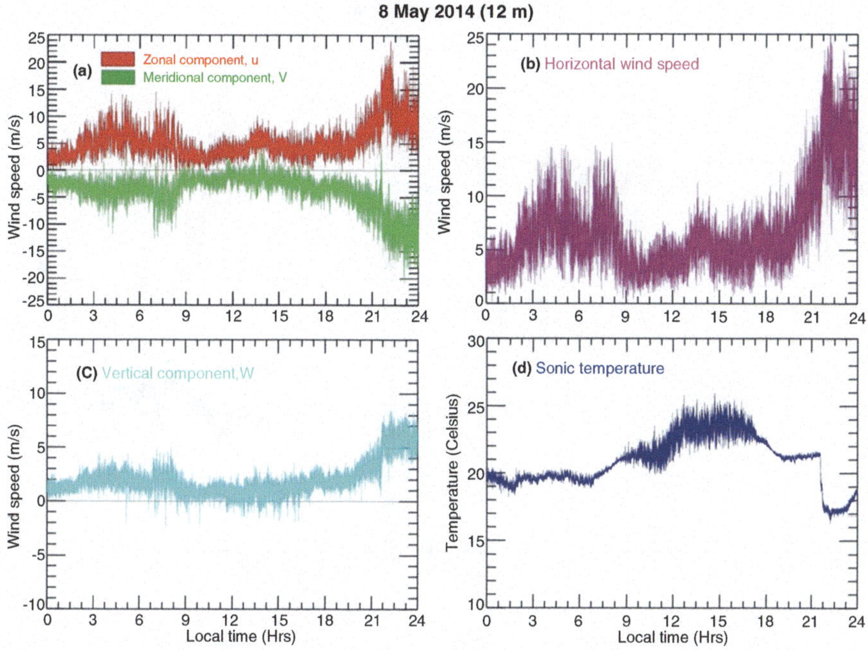

Fig. 3.6 Diurnal variability of horizontal wind components: zonal and meridional (**a**), horizontal wind speed (**b**), vertical wind component (**c**), and sonic temperature (**d**) on 8 May 2014 (12 m)

recorded had a horizontal wind speed of 24 ms^{-1} with vertical component as high as 8 ms^{-1}. This suggests that a strong vertical wind shear (gradient) has developed in the atmosphere which changed the precipitation pattern and injected cold moist air into the lower atmosphere. The significant meteorological parameters relating to this activity are presented in Fig. 3.6.

The above-mentioned fronts of this season belong to the category of cold fronts which affected the pre-monsoon weather. It is interesting to note that the intensity of cold fronts increased from March to May as the summer conditions set in.

3.4.3 Fronts Observed in Pre-Monsoon

Monsoon in the region generally reaches in early July; thus, the summer month of June can be considered as pre-monsoon with direct influence on monsoon circulation and its behavior. In the month of June 2014, two frontal activities were observed. The activity observed on 1 June was an interesting one which occurred in the morning hours with a sharp drop in temperature of 5°C. The horizontal wind acquired a speed of 24 $ms^{-1}$with a strong vertical component of 9 ms^{-1}. Later on from 10:00 hours (LT) to 14:00 hours (LT) the temperature rose again by 9°C and reached a maximum value of 24°C. A secondary drop in temperature by about 5°C was

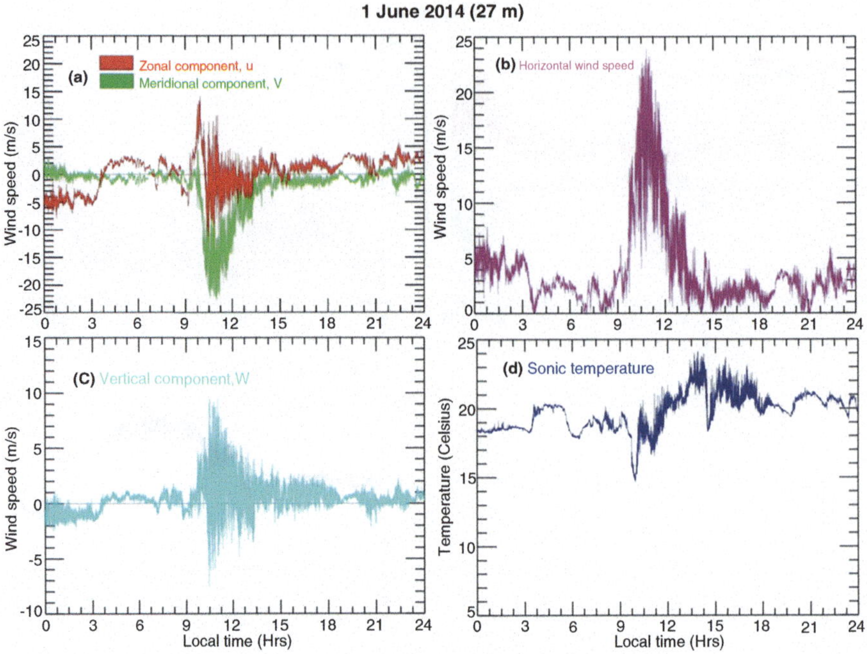

Fig. 3.7 Diurnal variability of horizontal wind components: zonal and meridional (**a**), horizontal wind speed (**b**), vertical wind component (**c**), and sonic temperature (**d**) on 1 June 2014 (27 m). Northerly wind component has reached as high as 24 ms^{-1} during the front

observed at 15:00 hours (LT). This phenomenon of an occluded front was active for the duration of almost 8 h. The wind and temperature characteristics observed during this frontal passage are presented in Fig. 3.7.

3.4.4 *Fronts Observed During Post-Monsoon*

During the months of July and August when monsoon activity is at its peak over the region, frontal activity gets pushed farther north and remains less active, however, immediately following the withdrawal of monsoon from the region in late September or early October. During this period, there is enough moisture available in the atmosphere and the temperatures are also moderately higher during the daytime. In 2014, a front was observed on 14 September at around 17:00 hours (LT) resulting into an initial temperature decrease followed by an increase of about 4° with horizontal wind speed of 17 ms^{-1} and an unusual high vertical wind speed of 5 ms^{-1}. Though the intensity of front was not very strong, the temperature fluctuation on either side from daily mean was quite substantial. The meteorological parameters associated with this occluded frontal system are depicted in Fig. 3.8.

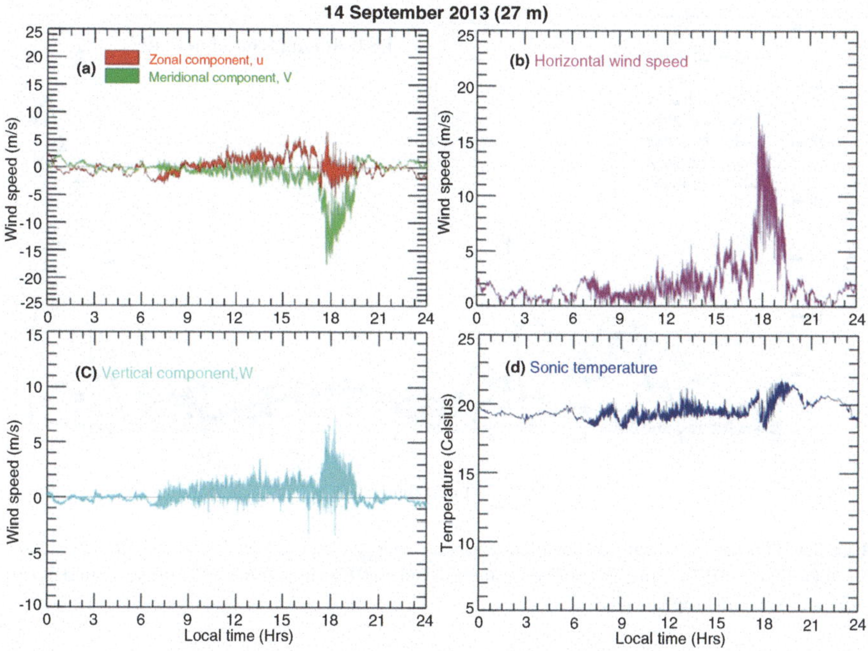

Fig. 3.8 Diurnal variability of horizontal wind components: zonal and meridional (**a**), horizontal wind speed (**b**), vertical wind component (**c**), and sonic temperature (**d**) on 14 September 2013 (27 m)

3.5 Turbulent Kinetic Energy Flux During Frontal Passage

The turbulence kinetic energy (TKE) in the atmospheric boundary layer is a measure of intensity of turbulence caused by unsteady wind fluctuations in the layers close to the ground. The TKE in simple terms can be understood as the kinetic energy of the transient wind components measured as departure from its time mean. In order to express the strength of the fronts discussed in the earlier sections in terms of their TKE, the values corresponding to individual frontal occurrences are plotted in a single diagram (Fig. 3.9). The presentation in the plot clearly brings out the diurnal variability of TKE during seven fronts observed over the site in 2013–2014. However, we have many more fronts observed over a period of the year, but only seven are presented here.

It is obviously demonstrated that whenever a front begins to burst there is the peak in TKE. The strength of the activity is reflected in sudden increase in TKE in all cases with largest magnitude of 35.0 $m^2\ s^{-2}$ during the episode of 1 June 2014. The front of 23 February 2013 shows the sharpest gradient in TKE. As a measure of magnitude of the frontal systems, it is worth noting that for all the occasions of frontal activity over the station, the mean TKE of non-frontal hours of all the events is less than 2.0 $m^2\ s^{-2}$.

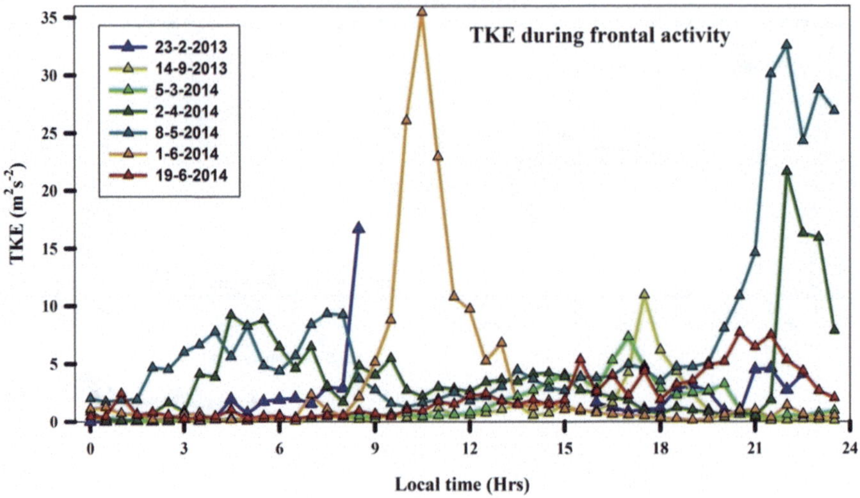

Fig. 3.9 Diurnal variability of turbulence kinetic energy observed during seven different fronts from February 2013 to June 2014. The front observed on 01 June witnesses the highest wind deviation and hence the highest TKE

References

Bond NA, Fleagle RG (1985) The structure of a cold front. Q J R Meteorol Soc 111:739–759

Dubey CS, Shukla DP, Ningreichon AS, Usham AL (2013) Orographic control of the Kedarnath disaster. Curr Sci 105(11):1474–1476

Guhathakurta P, Sreejith OP, Menon PA (2011) Impact of climate change on extreme rainfall events and flood risk in India. J Earth Syst Sci 120(3):359–373

Houze RA Jr (1993) Cloud dynamics. Academic, San Diego, CA. 337p

Jayaratne ER, Saunders CPR, Hallett J (1983) Laboratory studies of the charging of soft hail during ice crystals interactions. Q J R Meteorol Soc 109:609–630

Joshi V, Kumar K (2006) Extreme rainfall events and associated natural hazards in Alaknanda valley, Indian Himalayan region. J Mt Sci 3:228–236

Kumar PP, Saunders CPR (1989) Charger transfer during single crystal interaction with a rimed target. J Geophys Res 94:13099–13102

Nandargi S, Dhar ON (2011) Extreme rainfall events over the Himalayas between 1871 and 2007. Hydrol Sci J 56:930–945

Pant GB, Rupa Kumar K (1997) Climates of South Asia. John Wiley & Sons, New York/London. 320p

Reynolds SE, Brook M, Gourley MF (1957) Thunderstorm charge separation. J Meteor 14:426–436

Saunders CPR (2008) Charge separation mechanisms in clouds. Space Sci Rev 137:335–353

Schneider SH (1996) Encyclopedia of climate and. Weather 1:81–82

Schultz DM (2005) Review of cold fronts with prefrontal troughs and wind shifts. Mon Weather Rev 133:2449–2472. doi:10.1175/MWR2987.1

Schultz DM (2006) Perspectives of Fred Sanders' research on cold fronts. In: Synoptic–dynamic meteorology and weather analysis and forecasting: a tribute to Fred Sanders. Meteorological monographs, vol 55. American Meteorological Society, MA, pp 109–126

Solanki R, Singh N, Kiran Kumar NVP, Rajeev K, Dhaka SK (2016) Time variability of surface-layer characteristics over a mountain ridge in the Central Himalayas during the spring season. Bound-Lay Meteorol 158(3):453–471

Takahashi T (1978) Riming electrification as a charge generation mechanism in thunderstorms. J Atmos Sci 35:1536–1548

Tong HW, Chan JCL, Zhou W (2009) The role of MJO and mid-latitude fronts in the South China Sea summer monsoon onset. Clim Dyn 33:827–841. doi:10.1007/s00382-008-0490-7

Wallace JM, Hobbs PV (1977) Atmospheric science, an introductory survey. Academic Press, New York. 467p

Wilby R (1995) Simulation of precipitation by weather pattern and frontal analysis. J Hydrol 173(1–4):91–109

Chapter 4
Climate Models, Projections, and Scenarios

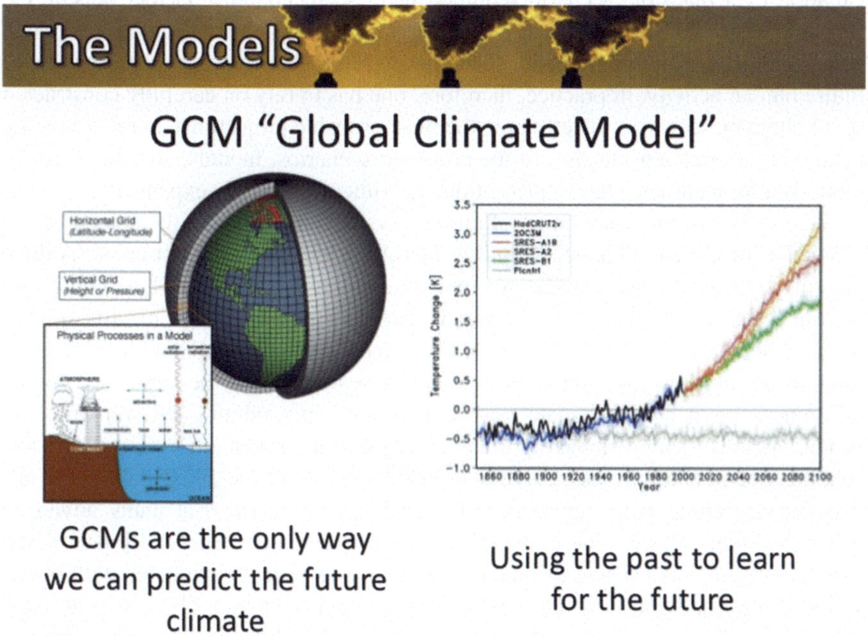

4.1 General Background

Hierarchy of climate models currently in use involve numerical representation of climate system based on known physical, chemical, and biological properties of their components, interactions, and feedbacks to generate future projections. Climate

G.B. Pant et al., *Climate Change in the Himalayas*,
DOI 10.1007/978-3-319-61654-4_4

projections are the descriptions of the model simulated response of the climate system to scenarios of greenhouse gases and aerosol concentrations or some hypothetical forcings on the climate components. The projections are used to generate scenarios of future climate for a specified time domain. The scenarios on the other hand are plausible and often simplified description of how the future climate may develop based on coherent and internally consistent set of assumptions about key driving forces and relationships. The climate scenarios often make use of projections through appropriate diagnostic model outputs and comparison with observed climate data. The scenarios are constructed for explicit use in investigating the potential impacts of anthropogenic factors on climate; therefore, they should not be confused with climate predictions. In principle, the climate models may be able to predict future climates on decadal to century scale so as to provide broad frame work for relevant policy decisions; however, the ability to do so have constraints because of limitations on precisely generating the probable emission scenarios of GHGs and other anthropogenic influences. Creation of a plausible scenario requires many socioeconomic factors outside the scope of basic science disciplines such as trend of change in population, economic policy, technological development, social norms, and other relevant characteristics of future human activity. In practice, therefore, one has to rely on carefully constructed set of alternate scenarios to determine the most probable projections. In the following paragraph, a brief introduction to the emission scenarios, models, and the status of input data for a climate change projection experiment are briefly explained.

Very nature of the functioning of climate system suggests that the detection and projection of climate change, including possible changes in the extremes, is done globally and specifically scaled down for examination in greater detail over the target region. IPCC assessments have indicated that the current versions of atmosphere-ocean general circulation models (AOGCMs) have generally well simulated the features of the present-day climate at continental scales though the large inter-model differences on a regional scale with consequent uncertainties still remain. The AOGCMs have demonstrated significant and rapid improvement over the past few years as documented in successive assessment reports of the IPCC during the last two decades along with the results of experiments performed at many advanced research centers all over the globe. However, it is essential that the climate model simulations are down scaled to smaller regions, systematically validated with observations, and finally suitable regional climate change scenarios along with reliable estimations of associated uncertainties are developed. With the increasing interest and rapid developments in the arena of global and regional modeling, during past few decades it is most likely that ready to use estimates of future projections on climate change scenarios may be available for wide spectrum of users in the next few decades.

Variety of experiments have been performed by different modeling groups across the world to simulate the expected climate change patterns under different emission scenarios prepared under IPCC coordination. Most commonly used scenarios are the IS92a (one of the six alternative IPCC scenarios published in 1992) and SRES (Special Report on Emissions Scenarios) as described in the IPCC Report (Nakicenovic and Swart 2000). These scenarios are fetched with extensive model simulated data made available by the IPCC Data Distribution Centre (IPCC-DDC) and the WCRP-CMIP3

(Coupled Model Intercomparison Project) data facility at the PCMDI (Program for Climate Model diagnosis and Intercomparison), San Francisco, to the research community working on climate change. Though the global atmosphere-ocean coupled models have provided good representations of the planetary scale features, but their application to regional studies is limited by their coarse resolution (~300 km). It is known on global climate modeling that very large computers operating continuously for very long time periods are required to create simulations of the earth's global climate. As more details of the climate system (e.g., ocean circulation, ocean-ice dynamics, vegetation, etc.) are included in climate models, this computational burden becomes even more severe. At the same time, the coarse resolution of global climate models does not allow them to simulate local climate conditions that may have huge impact, say, details regarding mountainous areas that are important or climate variations in the coastal regions. In fact, global climate models (GCMs) at low resolution do not include orographic details indicating that GCM runs are not sufficient or suitable for an in-depth study. Therefore, to overcome the resolution issue, RCMs are employed instead. For future climate change studies, however, the lateral forcing is to be obtained from a model such as a GCM. For example, these models do not contain realistic topographical features like the Western Ghats along the west coast of India, and consequently fail to reproduce their predominant influence on the peninsular monsoon rainfall patterns (Rupa Kumar and Ashrit 2001; Rupa Kumar et al. 2002, 2003). Developing high resolution models on a global scale is not only computationally expensive for climate change simulations, but also one has to deal with the possible errors due to inadequate representation of high resolution climate processes on global scale.

The RCMs provide an opportunity to dynamically downscale global model simulations to superimpose the detail of the specific regions. Development of high resolution (much smaller grid size than the one used in global models) climate change scenarios may prove to be very useful in the following ways: (a) a realistic simulation of the current climate by taking into account fine-scale features of the terrain, (b) more detailed predictions of future climate changes taking into account local weather features and responses, (c) representation of the smaller islands and their unique features, (d) better simulation and prediction of extreme climatic events, and (e) generation of detailed regional data to drive other region-specific models analyzing local-scale impacts (Noguer et al. 2002). Increasingly reliable regional climate change projections are now available for many regions of the world but the limitations due to uncertainties in the factors such as population change, economic variables, technological developments, and other relevant characteristics of future human activities remain. Therefore, certain carefully considered scenarios of plausible future socioeconomic pathways are designed and used to estimate the associated global climatic scenarios for the future and to assess their likely consequences. Most of the climate change impact studies so far have been focused on the mean climate; however, with initial promise demonstrated by modeling experiments it is encouraging to note that the reliability on the prediction of variability and extremes of weather and climate have substantially improved in the last few decades.

Use of RCM (Marinucci and Giorgi 1992; Machenhaueret al. 2001; Bell et al. 2004) driven by large-scale forcings coming from Global Climate models that have been proved in the recent years is one of the most interesting and fruitful procedures to study regional climate changes.

4.2 Advantages and Disadvantages of RCMs

With improvement in the numerical weather forecasting capabilities, various schemes are being developed and extensively used at global as well as regional scales, depending upon the application and utility in order to furnish better forecast. Weather research and forecast (WRF) model is being popularly used at regional scale for meteorology and chemistry as well. With great advantages, the regional climate models have some inherent disadvantages as well. Regional climate models can be used to study specific areas in more detail as RCMs cover smaller areas, i.e., they can have higher spatial resolution and the denser grid gives better results. Over a global model, RCM has an advantage to include the ability to study the natural variability and it can also estimate possible future climate change.

However, the inherent disadvantage to the RCM is that because of computational limitations imposed by the higher resolution and much shorter time steps therefore required, they can only be run for a limited area domain. Regional models, because they do not span the entire globe, must rely on information provided at the lateral boundaries in order to simulate climates for the interior of their model domains. Moreover, regional models don't replace global models but might be able to supply added value to simulations done with global models and will not correct for mistakes of GCM. These models have limited applications, i.e., run with few GCMs and often run for limited time periods.

4.3 Description on Scenarios

Scenarios are plausible combinations of conditions that can represent possible future situations, thus are images of the future, or alternative futures. They are neither predictions nor forecasts. Rather, each scenario is one alternative image of how the future might unfold. Scenarios are often used to assess the consequences of possible future conditions. Scenarios help in the assessment of future developments in complex systems that are either inherently unpredictable, or that have high scientific uncertainties. In all stages of the scenario-building process, uncertainties of different nature are encountered. Climate change scenarios are scenarios of plausible changes in climate. We use them to understand what the consequences of climate change can be. We can also use them to identify and evaluate adaptation strategies and aid the assessment of future climate change, impacts, vulnerabilities, adaptation, and mitigation.

It is critical to keep in mind that regional climate change scenarios are not a prediction of future climate change, but rather a tool to communicate what could happen as a result of human-induced climate change and to facilitate understanding of how different systems could be affected by climate change.

4.4 Special Report on Emissions Scenarios (SRES)

It is important to remember that future scenario calculations are based on assumptions on the future world development. A good quality climate model may provide a plausible realization of the weather with all statistical properties but may not be able to reproduce precise weather conditions in space and time.

Because projections of climate change depend heavily upon future human activity, climate models are run against scenarios. There are 40 different scenarios, each making different assumptions for future greenhouse gas pollution, land use, and other driving forces. Assumptions about future technological development as well as the future economic development are thus made for each scenario.

These emission scenarios are organized into families, which contain scenarios that are similar to each other in some respects. IPCC assessment report projections for the future are often made in the context of a specific scenario family.

4.4.1 A1 Scenarios

The A1 scenarios are of a more integrated world. The A1 family of scenarios is characterized by:

1. Rapid economic growth.
2. A global population that reaches nine billion in 2050 and then gradually declines.
3. The quick spread of new and efficient technologies.
4. A convergent world—income and way of life converge between regions, extensive social and cultural interactions worldwide.

4.4.1.1 Technological Emphasis

There are subsets to the A1 family based on their technological emphasis:

1. A1FI—An emphasis on fossil-fuels
2. A1B—A balanced emphasis on all energy sources
3. A1T—Emphasis on nonfossil energy sources

4.4.2 A2 Scenarios

The A2 scenarios are of a more divided world. The A2 family of scenarios is characterized by:

1. A world of independently operating, self-reliant nations
2. Continuously increasing population

3. Regionally oriented economic development
4. Slower and more fragmented technological changes and improvements of per capita income

4.4.3 B1 Scenarios

The B1 scenarios are of a world more integrated and more ecologically friendly. The B1 scenarios are characterized by:

1. Rapid economic growth as in A1, but with rapid changes towards a service and information economy
2. Population rising to nine billion in 2050 and then declining as in A1
3. Reductions in material intensity and the introduction of clean and resource efficient technologies
4. An emphasis on global solutions to economic, social and environmental stability

4.4.4 B2 Scenarios

The B2 scenarios are of a world more divided, but more ecologically friendly. The B2 scenarios are characterized by:

1. Continuously increasing population, but at a slower rate than in A2
2. Emphasis on local rather than global solutions to economic, social, and environmental stability
3. Intermediate levels of economic development
4. Less rapid and more fragmented technological change than in A1 and B1

In a recent study especially designed to simulate the regional climate over Central Himalayan region, a high resolution regional climate model Providing REgional Climate for Impact Studies (PRECIS) developed by the Hedley Centre of the United Kingdom Meteorological Office was performed by the Indian Institute of Tropical Meteorology, Pune. Data for three perturbed physics ensembles under the scheme of Quantifying Uncertainties in Model Prediction (QUMP) are used (Rupa Kumar et al. 2006; Krishna Kumar et al. 2011). For the purpose of generating future climate change scenarios on decadal scales over Central Himalaya (CH), an appropriate grid box (74.5–84.5°E and 25.5–34.0°N) is selected. This box representing the central Himalaya in the model domain covers the hilly parts of Himachal Pradesh, Uttarakhand, and adjoining western Nepal. For reliable model simulations, it is most essential that the best possible initial data are generated based on observations for each grid point in the model domain and examined for their representativeness.

4.5 Representative Concentration Pathways (RCPs)

SRES was superseded by Representative Concentration Pathways (RCPs) in 2014. In preparation for the Fifth Assessment Report (AR5) for IPCC, researchers developed a new approach for creating and using scenarios in climate change research. The new approach is built around the concept of Representative Concentration Pathways (RCPs). RCPs are time- and space-dependent trajectories of concentrations of greenhouse gases and pollutants resulting from human activities and including changes in land use. RCPs provide a quantitative description of concentrations of the gases and one RCP provides only one of many possible scenarios that would lead to the specific radiative forcing pathway. Radiative forcing plays an important in climate change and is a measure of the additional energy taken up by the Earth system due to increased pollutants in the atmosphere over time. RCPs are used in current scenarios as well as their radiative forcing by 2100 in climate models. Total radiative forcing is determined through both positive forcing from greenhouse gases and negative forcing from aerosols. The dominant factor by far is the positive forcing from CO_2. As the radiative forcing increases the global temperature rises.

The pathways are used for climate modeling and research the following four RCPs were selected and defined by their total radiative forcing (mainly cumulative measure of human emissions of GHGs from all sources expressed in W m^{-2}) and level by 2100 (IPCC AR5). The RCPs were chosen to represent a broad range of climate outcomes, based on extensive literature review, and are neither forecasts nor policy recommendations.

4.5.1 *RCP 8.5: High Emissions*

The RCP 8.5 is characterized by increasing greenhouse gas emissions over time and a representative for scenarios leading to high greenhouse gas concentration levels. Rising radiative forcing pathway will be leading to 8.5 W m^{-2} in 2100 (Riahi et al. 2007 and Rao and Riahi 2006).

4.5.2 *RCP 6: Intermediate Emissions*

It is a stabilization scenario where total radiative forcing is stabilized after 2100 without overshoot by the employment of a range of technologies and strategies for reducing greenhouse gas emissions. Stabilization without overshoot pathway is demonstrated to be 6 Wm^{-2} after 2100 (Fujino et al. 2006 and Hijioka et al. 2008).

4.5.3 RCP 4.5: Intermediate Emissions

It is another important stabilization scenario where total radiative forcing is stabilized before 2100 by employment of a range of technologies and strategies to be used, in order to reduce the greenhouse gas emissions. Stabilization without overshoot pathway turns out to be to 4.5 Wm^{-2} at stabilization after 2100 (Smith and Wigley 2006; Clarke et al. 2007; Wise et al. 2009).

4.5.4 RCP 2.6: Low Emissions

The emission pathway is representative for scenarios leading to very low greenhouse gas concentration levels. It is the so-called peak scenario and its radiative forcing level first reaches a value around 3.1 Wm^{-2} in mid-century, and finally returning to 2.6 Wm^{-2} by 2100. Peak in radiative forcing at ~3 Wm^{-2} before 2100 and subsequently followed by a decline (van Vuuren et al. 2006, 2007).

4.6 Overview of Past Studies

4.6.1 Global Context

As the climate changes, extreme weather events such as meteorological droughts, floods, heat and cold wave, heavy rain spells and cyclones are likely to show some change in their behavior. Quite often the patterns of persistent extreme weather may lead to an extreme climate event which will form an important aspect of climate change. Changes in the frequency as well as intensity of extreme climate events would have profound impacts on human society, infrastructure, natural resources, and ecosystem more significantly on regional scale. With growing concern towards the regional manifestations of global warming, one of the important aspects of climate change research is the quantitative detection of precipitation and temperature extremes, their spatial and temporal variability and future projections.

In a global context, the model simulations have arrived at many coherent conclusions, the most prominent among them being that all land regions are likely/very likely to warm in the twenty-first century (IPCC 2007). Interannual temperature variability is likely to increase in summer in most areas (Giorgi and Bi 2005; Rowell 2005; Clark et al. 2006; Schär et al. 2004; Vidale et al. 2007). Heat waves are likely to increase in frequency, intensity, and duration along with overall warming and changes in variability (Barnett et al. 2006; Clark et al. 2006; Tebaldi et al. 2006). Tebaldi et al. (2006) have also shown that number of frost days is very likely to decrease. Most of the models show increase in extreme daily precipitation despite decrease in mean precipitation. Much larger changes are expected in the recurrence frequency of precipitation extremes than in the magnitude of extremes (Huntingford et al. 2003; Barnett et al. 2006; Frei et al. 2006). Decrease in number of precipitation days and increase in the length of the longest dry spell in southern and central Europe has been indicated by several models (Semenov and Bengtsson 2002; Voss et al. 2002; Räisänen 2003, 2004; Pal et al. 2004; Frei et al. 2006; Beniston et al. 2007; Gao et al. 2006; Tebaldi et al. 2006). The latest fifth assessment report of the Intergovernmental Panel on Climate Change (IPCC-WG-1 2013) concluded that the global mean surface temperatures have risen by 0.78 °C (0.72–0.85 °C) when estimated trend over the 1850–1900 to 2003–2012 most recent decade.

4.6.2 Asian Region

Over Asia warming is likely to be well above the global mean and precipitation is likely to increase over parts of the Asia. It is likely that heat waves/hot spells in summer will be of longer duration, more intense and more frequent and fewer very cold days in East Asia. Frequency of intense precipitation events, and also extreme rainfall and winds associated with tropical cyclones are likely to increase in East Asia, Southeast Asia, and South Asia (IPCC 2007). Though monsoonal flows and tropical large-scale circulation weaken (Knutson and Manabe 1995), enhanced moisture convergence in a warmer, moister atmosphere dominates over it and result into increased monsoon precipitation (Douville et al. 2000; Giorgi et al. 2001a, b; Stephenson et al. 2001; Dairaku and Emori 2006; Ueda et al. 2006). The pattern of ocean temperature change across Pacific is of central importance to climate change over Asia. Interannual rainfall variability is significantly affected in Asia over central and Southeast Asia (McBride et al. 2003).

Decrease in DJF precipitation and increase in rest of the year has been projected by most of the models in the multi-model data set for the A1B scenario (MMD-A1B) considered by the IPCC Fourth Assessment Report (IPCC 2007). Increased precipitation for most of the year was also seen in the earlier AOGCM simulations (Lal and Harasawa 2001; Lal et al. 2001; Rupa Kumar and Ashrit 2001; Rupa Kumar et al. 2002, 2003; Ashrit et al. 2003; May 2004). A general increase in the intensity of heavy rainfall events in future over north-west India as well as north-east India is seen by May (2004). Increase in the precipitation intensity as well as extreme precipitation is simulated under the IS92a scenario in the 2050s (Krishna Kumar et al. 2003). Projection of increase in extreme precipitation is mainly due to the dynamic effect of enhanced upward motion due to northward shift of monsoon circulation, as seen through an AGCM by Dairaku and Emori (2006). Using India Meteorological Department (IMD) temperature data at a resolution of $1° \times 1°$, Oza and Kishtawal (2015) have shown that India has dominance tendency towards the warming and higher warming during winter months of the year. Their analysis indicates south and west India generally show rising trend mentioned by Sinha Ray and De (2003).

Beniston and Rebetez (1996) attribute their finding of greater warming at higher altitudes to the fact that high-elevation stations are more directly in contact with the free troposphere than low-elevation ones, and therefore, less affected by increasing anthropogenic factors such as urbanization and pollution. However, study made by

Revadekar et al. (2013) on south Asian region mentioned that high-elevation sites appear to be more influenced by local factors.

An increase is observed in the frequency and intensity of weather events over Koshi river basin which is a subbasin of the Ganges shared among China, Nepal, and India (Shrestha et al. 2016). Caesar et al. (2011) considering seven stations from Nepal showed that warm extremes are increasing and cold extremes are decreasing. Using CMIP5 data, Murari et al., (Murari et al. 2015) presented projections of future heat waves in India based on multiple climate models and scenarios. Their study shows that heat waves are expected to intensify around the globe in the future, with potential increase in heat stress and heat-induced mortality in the absence of adaptation measures. It has been projected that the global mean surface temperature and sea level may increase by 0.3–1.7 °C and 0.26–0.54 m for RCP2.6, 1.1–2.6 °C and 0.32–0.62 m for RCP4.5, 1.4–3.1 °C and 0.33–0.62 m for RCP6.0, and 2.6–4.8 °C and 0.45–0.81 m for RCP8.5, respectively, by 2081–2100 (IPCC 2013). Temperature and rainfall changes have also been reported by other studies made by Jones and Briffa (1992), Guhathakurta and Rajeevan (2008), Tyagi and Goswami (2009), Kothawale et al. (2010). Turner and Annamalai (2012) have described the monsoon and the impact of changing climate on South Asia. Changes in south-west monsoon rainfall over India have also been studied by Guhathakurta et al. 2015. Guhathakurta et al. 2011 analyzed the impact of climate changes on extreme rainfall events and flood risk in India. Recent study of Srivastava et al. revealed that annual mean, maximum, and minimum temperatures averaged over the country as a whole exhibited a significant increasing trend of 0.60 °C, 1.0 °C, and 0.18 °C per hundred years, respectively.

4.6.3 PRECIS (Providing REgional Climates for Impacts Studies) Scenarios

PRECIS stands for Providing REgional Climates for Impacts Studies, and it is a third-generation regional climate modeling system with a very flexible design that allows it to be applied in any region of the globe. It is also driven by the boundary conditions like any other regional climate model does, and boundary conditions are simulated by general circulation models. PRECIS has been successfully set in Europe, Indian subcontinent, and South Africa during its development and output also used in impact assessment (Jones et al. 2004). Rupa Kumar et al. (2006) have shown that PRECIS simulations under scenarios of increasing greenhouse gas concentrations and sulfate aerosols indicate marked increase in both rainfall and temperature over India towards the end of twenty-first century. Surface air temperature and rainfall show similar patterns of projected changes under A2 and B2 scenarios, but the B2 scenario shows slightly lower magnitudes of the projected change. Temperatures are likely to increase in entire calendar year, but the changes in winter season are expected to be prominent. Diurnal temperature range is expected to decrease in winter (JF) and pre-monsoon (MAM) months (Revadekar et al. 2012).

The warming is monotonously widespread over the country, but there are substantial spatial differences in the projected rainfall changes. West-central India shows maximum expected increase in rainfall. Extreme precipitation shows substantial increases over large area, particularly over the west coast of India and west-central India. However, their study does not explore on the seasonal changes in extremes. General warming and enhanced rainfall over India is indicated by coupled atmosphere-ocean general circulation models also in a greenhouse increase scenarios (Lal and Harasawa 2001; Lal and Singh 1998; Rupa Kumar and Ashrit 2001; Rupa Kumar et al. 2002, 2003).

Analysis on frequency and intensity of cyclonic disturbances during the monsoon seasons and their likely changes in future has been examined by Patwardhan et al. (2008). Their study indicates that the frequency of cyclonic disturbances is likely to decrease in future; however, the systems may be more intense in warming scenarios. The study also indicates that there is no significant change in the onset of monsoon over Kerala in future scenarios, but a high variability exists in A2 scenarios as compared to the baseline. The analysis done by Nazrul Islam (Nazrul Islam and Uyeda 2007) also reveals similar results on Bangladesh. The rate of increase in rainfall will vary from 1 to 2.5 mm per day, whereas that for temperature will vary from 2.1 to 3.4 °C. Over China, Yinlong Xu et al. (2006) have demonstrated that the future extreme maximum temperature and precipitation events would increase, while future extreme minimum temperature events would decrease during 2071–2100 under B2 scenarios. Their study also indicated that both the flooding in summer and drought in winter would enhance over central, east, and south China. RCM (PRECIS) also project substantial increases in extreme precipitation towards the end of twenty-first century over large area, particularly over the west coast of India and west-central India (Rupa Kumar et al. 2006).

4.6.4 Himalayan Climate in Changing Scenarios

Observed data indicated that the rate of warming in Central Himalayan region is significantly higher than the global average of 0.74 °C over the last 100 years. Within the region, the rates for the Central Himalaya (Nepal) and Tibetan plateau (based on limited data) appear to be considerably higher (0.04–0.019 °C/year and 0.03–0.07 °C/year, respectively). The measurements in Nepal and Tibet also indicate that warming is occurring at much higher rates in the high altitude regions than in low attitudes (Shrestha et al. 1999; Liu and Chen 2000). The total rainfall does not show any distinct and consistent trend in these regions including the Indian part of Central Himalaya. In spite of the fact that the region has limited data, the large coherency among the data sets of different subregions due to the dominance of large-scale weather systems and their modulations due to climate broadly indicate the significance of Central Himalaya as one unit.

4.6.4.1 PRECIS Scenarios over Himalaya

The PRECIS simulations under scenarios of increasing greenhouse gas concentrations and sulfate aerosols indicate marked increase in both rainfall and surface air temperature over India towards the end of twenty-first century. Surface air temperature and rainfall show similar pattern of projected changes under A2 and B2 scenarios, but the magnitude of change under B2 scenario is slightly lower than A2 (Rupa Kumar et al. 2006). The warming is monotonously widespread over entire country, where as there are substantial spatial differences in the projected rainfall changes. West-central India shows maximum expected increase in rainfall. Extremes in maximum and minimum temperatures also indicate an increasing trend, with night temperatures increasing faster than the day temperatures. Extreme precipitation shows substantial increases over large area, particularly over the west coast of India and west-central India, though the seasonal changes in extremes are not explored.

As far as Himalayan region is concerned, hottest day temperature and coldest night temperature are likely to increase by 4–8 °C during winter DJF as can be seen in Fig. 4.1. Warming of the coldest night temperature is observed over a large area

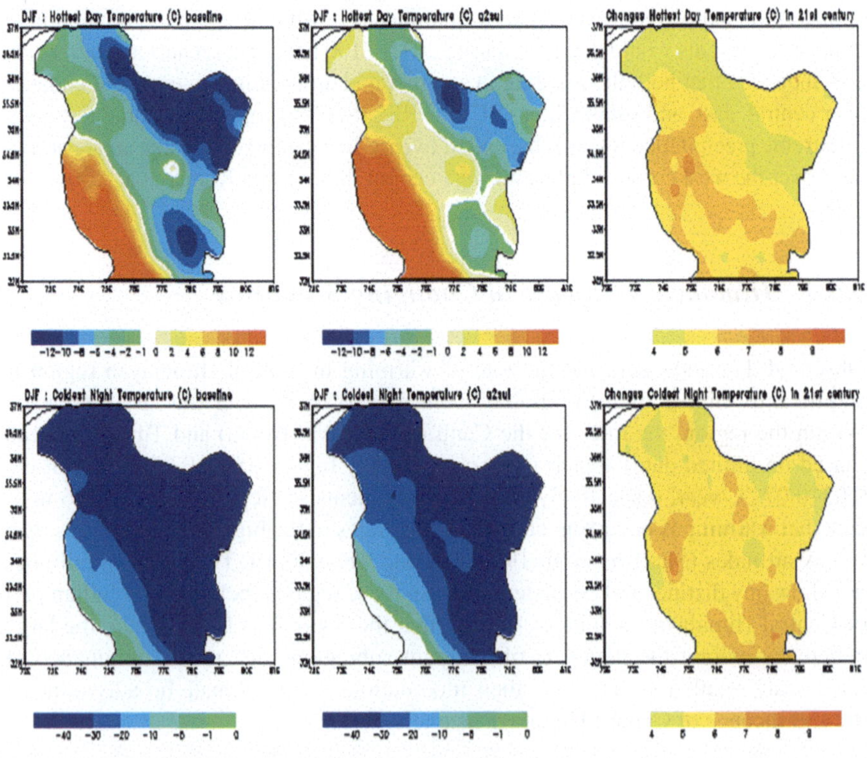

Fig. 4.1 Hottest day temperature (*top*) and coldest night temperature (*bottom*) over WH as simulated by PRECIS. Baseline, 1961–1990 (*left*), Scenarios 2071–2100, (*middle*), and change calculated as scenarios minus baseline (*right*)

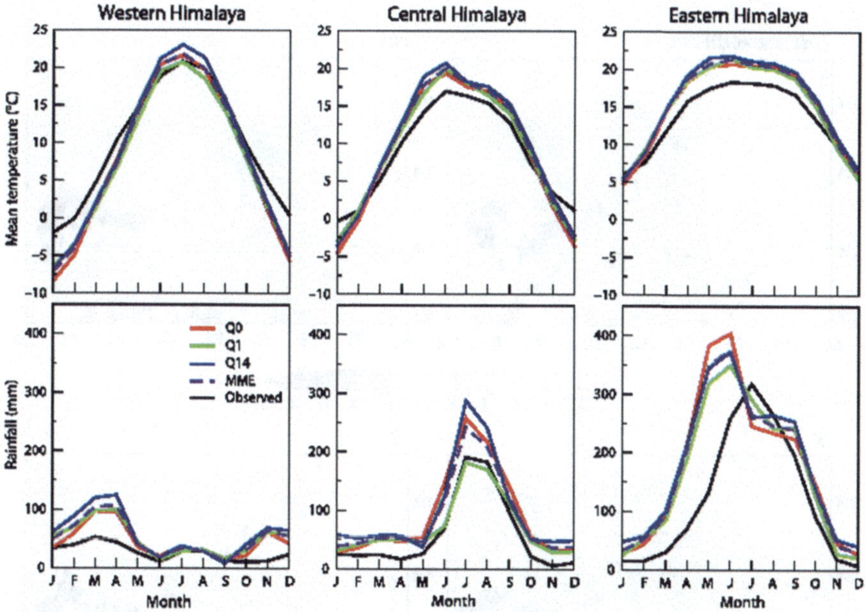

Fig. 4.2 Annual cycles of surface air temperature (*top*) and rainfall (*bottom*) over Western Himalaya, Central Himalaya, and Eastern Himalaya stimulated by three QUMP experiments Q0 (*red*), Q1 (*green*), Q14 (*blue*), their MME (*magenta dashed line*), and the observed APHRODITE rainfall and NCEP Reanalysis temperature (*black*) based on 1961–1990. (Kulkarni et al. 2013)

of central Himalaya. Ashwini Kulkarni et al. (2013) examined the potential impact of global warming on the mountainous regions of Hindu Kush Himalayan region by applying Hadley Centers high resolution regional climate model PRECIS to three subregions: 1-The Western, 2-Central, and 3-Eastern Himalayan. Annual cycles based on the monthly mean temperature and rainfall (1961–1990) simulated by the three QUMP simulations (Q0, Q1, and Q14) over central Himalaya are shown in Fig. 4.2. Observed rainfall is taken from APHRODITE and temperature from the NCEP reanalysis. Shapes of annual cycles of both temperature and rainfall are captured reasonably well. Summer monsoon (JJAS) temperatures are well simulated however winter season show cold bias. Summer monsoon peak rainfall is also well simulated by the models rainfall is overestimated.

The spatial patterns of the summer monsoon rainfall over Western Himalaya as simulated by the three QUMP runs for the baseline period 1961–1990 were compared with the observed APHRODITE precipitation patterns in Fig. 4.3. Present-day seasonal spatial rainfall show adequate representation by PRECIS. Wet bias is seen over northern parts, but Q1 show dry bias. Spatial patterns of annual average

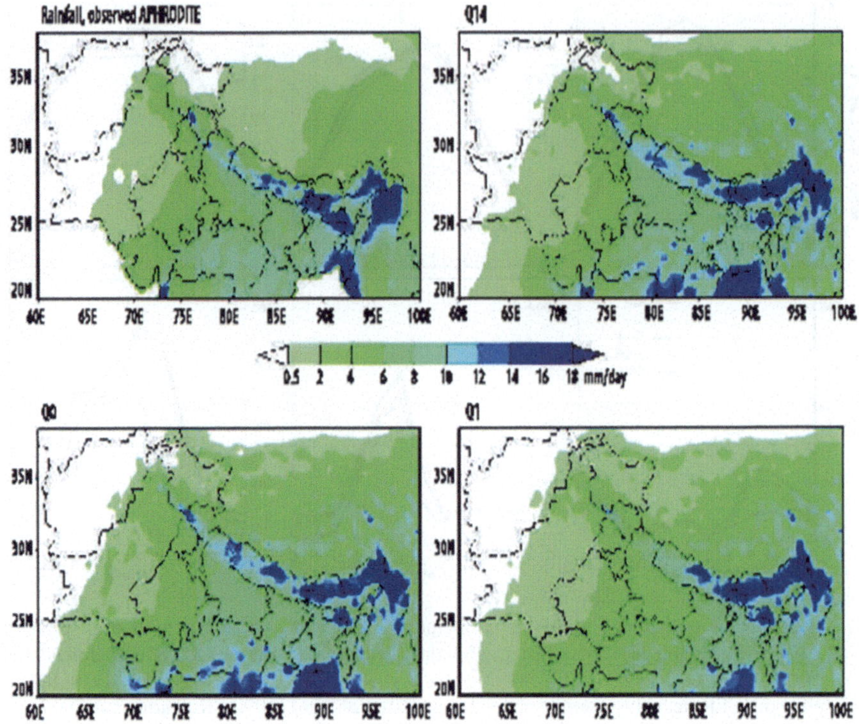

Fig. 4.3 Composite seasonal (June to September) monsoon rainfall (mm/day) simulated by three QUMP runs compared with the observed APHRODITE values for 1961–1990 (Kulkarni et al. 2013)

temperature are shown in Fig. 4.4 for 1961–1990. Temperature gradients over western Himalaya are well captured by the models with an overestimation over northern parts.

Model projections over Western Himalaya are examined for seasonal rainfall and annual surface temperature over the short (2011–2020), medium (2021–2050), and long term (2051–2080) in Fig. 4.5. Summer monsoon precipitation is expected to be 20–40% higher in 2051–2080 than it was in the baseline period (1961–1990). The model projections indicate that the significant warmings is likely to take place over the region towards end of the twenty-first century. The three Quantifying Uncertainties in Model Predictions simulations show large differences in projections in the western Himalaya.

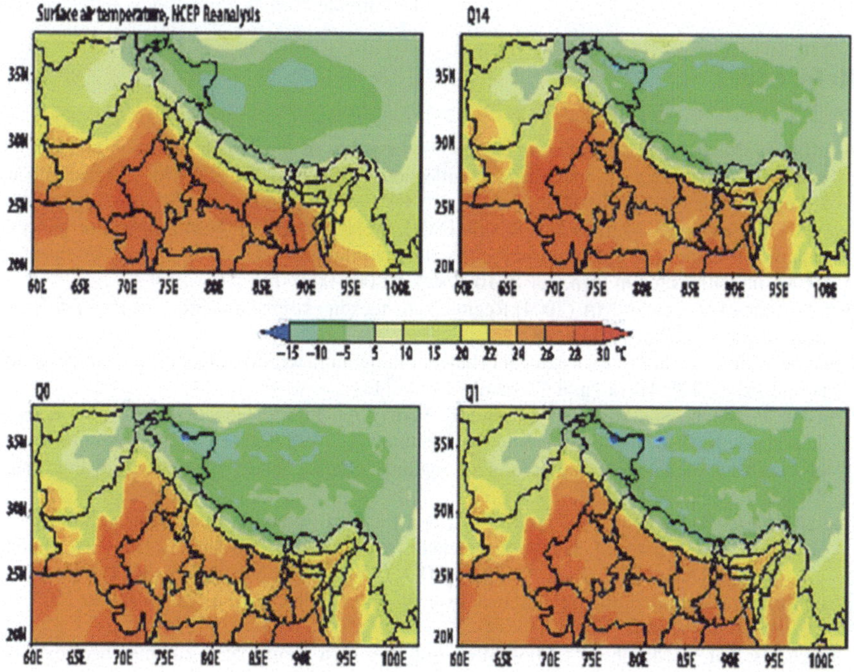

Fig. 4.4 Composite annual average surface air temperature (°C) simulated by the three QUMP runs compared with observed NCEP Reanalysis values for 1961–1990 (Kulkarni et al. 2013)

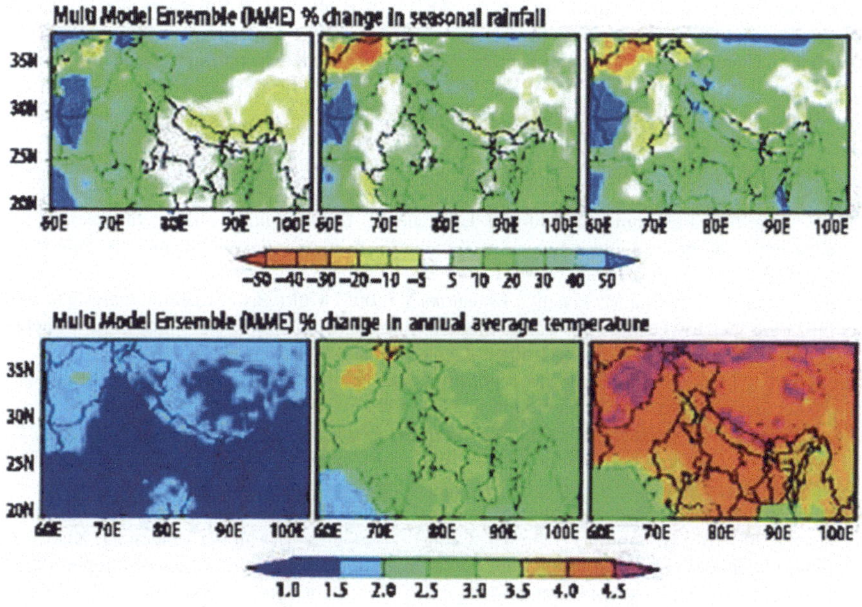

Fig. 4.5 MME average change in seasonal rainfall (*top*) and change in annual average temperature (*bottom*) in three time slices 2020s (*left*), 2050s (*middle*), and 2080s (*right*) with respect to baseline 1970 (Kulkarni et al. 2013)

References

Ashrit RG, Douville H, Rupa Kumar K (2003) Response of the Indian monsoon and ENSO-monsoon teleconnection to enhanced greenhouse effect in the CNRM coupled model. J Meteorol Soc Jpn 81:779–803

Barnett DN, Brown SJ, Murphy JM, Sexton DMH, Webb MJ (2006) Quantifying uncertainty in changes in extreme event frequency in response to doubled CO_2 using a large ensemble of GCM simulations. Clim Dyn. doi:10.1007/s00382-005-0097-1

Bell JL, Sloan LC, Snyder MA (2004) Regional changes in extreme climatic events: a future climate scenario. J Clim 17(1):81–87

Beniston M, Rebetez M (1996) Regional behavior of minimum temperatures in Switzerland for the period 1979-1993. Theor Appl Climatol 53:231–243

Beniston M, Stephenson DB, Christensen OB, Ferro CAT, Frei C, Goyette S, Halsnaes K, Holt T, Jylhä K, Koffi B, Palutikof J, Schöll R, Semmler T, Woth K (2007) Future extreme events in European climate; an exploration of regional climate model projections. Clim Chang 81(2007):71–95

Caesar J, Alexande LV, Trewin B, Tse-ring K, Sorany L, Vuniyayawa V, Keosavang N, Shimana A, Htay MM, Karmacharya J, Jayasinghearachchi DA, Sakkamart J, Soares E, Hung LT, Thuong LT, Hue CT, Dung NTT, Hung PV, Cuong HD, Cuongo NM, Sirabaha S (2011) Changes in temperature and precipitation extremes over the Indo-Pacific region from 1971 to 2005. Int J Climatol 31:791–801

Clark R, Brown S, Murphy J (2006) Modelling Northern Hemisphere summer heat extreme changes and their uncertainties using a physics ensemble of climate sensitivity experiments. J Clim 19:4418–4435

Clarke L, Edmonds J, Jacoby H, Pitcher H, Reilly J, Richels R (2007) Scenarios of greenhouse gas emissions and atmospheric concentrations. Sub-report 2.1A of synthesis and assessment product 2.1 by the U.S. Climate Change Science Program and the Subcommittee on Global Change Research. Department of Energy, Office of Biological & Environmental Research, Washington, DC. 154p

Dairaku K, Emori S (2006) Dynamic and thermodynamic influences on intensified daily rainfall during the Asian summer monsoon under doubled atmospheric CO_2 conditions. Geophys Res Lett 33. doi:10.1029/2005GL024754. ISSN: 0094-8276

Douville H, Viterbo P, Mahfouf JF, Beljaars ACM (2000) Evaluation of the optimum interpolation and nudging techniques for soil moisture analysis using FIFE data. Mon Weather Rev 128:1733–1756

Frei C, Schöll R, Fukutome S, Schmidli J, Vidale PL (2006) Future change of precipitation extremes in Europe: an intercomparison of scenarios from regional climate models. J Geophys Res 110(D3):4124–4137

Fujino J, Nair R, Kainuma M, Masui T, Matsuoka Y (2006) Multi-gas mitigation analysis on stabilization scenarios using AIM global model. Multigas Mitigation and Climate Policy. Energy J 3(Special issue). doi:10.5547/issn0195-6574-ej-volsi2006-nosi3-17

Gao X, Xu Y, Zhao Z et al (2006) Impacts of horizontal resolution and topography on the numerical simulation of East Asian precipitation. Chin J Atmos Sci 30:185–192. (in Chinese)

Giorgi F, Bi X (2005) Regional changes in surface climate interannual variability for the 21st century from ensembles of global model simulations. Geophys Res Lett 32:L13701. doi:10.1029/2005GL023002

Giorgi F et al (2001a) Regional climate information — evaluation and projections in climate change 2001. In: Houghton JT, Ding Y, Griggs DJ, Noguer M, van der Linden PJ, Xiaoxu D (eds) The scientific basis, contribution of working group I to the third assessment report of the Intergovernmental Panel on Climate Change (IPCC). Cambridge University Press, Cambridge, pp 583–638. (Chapter 10)

Giorgi F, Whetton PW, Jones RG, Christensen JH, Mearns LO, Hewitson B, von Storch H, Francisco R, Jack C (2001b) Emerging patterns of simulated regional climatic changes for the 21st century due to anthropogenic forcings. Geophys Res Lett 28:3317–3320

Guhathakurta P, Rajeevan M (2008) Trends in the rainfall pattern over India. Int J Climatol 28:1453–1469

Guhathakurta P, Sreejith OP, Menon PA (2011) Impact of climate change on extreme rainfall events and flood risk in India. J Earth Syst Sci 120(3):359–373

Guhathakurta P, Rajeevan M, Sikka DR, Tyagi A (2015) Observed changes in southwest monsoon rainfall over India during 1901–2011. Int J Climatol 35:1881–1898

Hijioka Y, Matsuoka Y, Nishimoto H, Masui M, Kainuma M (2008) Global GHG emissions scenarios under GHG concentration stabilization targets. J Glob Environ Eng 13:97–108

Huntingford C, Jones RG, Prudhomme C, Lamb R, Gash JHC (2003) Regional climate model predictions of extreme rainfall for a changing climate. Q J R Meteorol Soc 129:1607–1621

IPCC-WG-1 (2013) Climate change 2013. In: Stocker TF, Quin GK et al (eds) The physical science basis. Cambridge University Press, Cambridge. 1535p

Jones PD, Briffa KR (1992) Global surface air temperature variations during the 20th century: Part 1. Spatial, temporal and seasonal details. The Holocene 2:165–179

Jones RG, Noguer M, Hassell DC, Hudson D, Wilson SS, Jenkins GJ, Mitchell JFB (2004) Generating high resolution climate change scenarios using PRECIS. Met Office Hadley Centre, Exeter

Knutson TR, Manabe S (1995) Time-mean response over the tropical Pacific to increased CO_2 in a coupled ocean-atmosphere model. J Clim 8:2181–2199

Kothawale DR, Munot AA, Krishna Kumar K (2010) Surface air temperature variability over India during 1901–2007, and its association with ENSO. Clim Res 42:89–104

Krishna Kumar K et al (2003) Future scenarios of extreme rainfall and temperature over India. In: Proceedings of the workshop on scenarios and future emissions, Indian Institute of Management (IIM), Ahmedabad, July 22, 2003. NATCOM Project Management Cell, Ministry of Environment and Forests, Government of India, New Delhi, pp 56–68

Krishna Kumar K, Patwardhan SK, Kulkarni AK, Kamla K, Koteswar Rao K, Johns R (2011) Simulation and future projections of summer monsoon climate over India by a high resolution regional climate model (PRECIS). Curr Sci 101:312–326

Kulkarni A, Patwardhan S, Krishna Kumar K, Ashok K, Krishnan R (2013) Projected climate change in the Hindu Kush–Himalayan region by using the high-resolution regional climate model PRECIS. Mt Res Dev 33(2):142–151

Lal M, Harasawa H (2001) Future climate change scenarios for Asia as inferred from selected coupled atmosphere-ocean global climate models. J Meteorol Soc Jpn 79(1):219–227

Lal M, Singh SK (1998) Global warming and monsoon climate. Mausam 52:245–262

Lal M, Nozawa T, Emori S, Harasawa H, Takahashi K, Kimoto M, Abe-Ouchi A, Nakajima T, Takemura T, Numaguti A (2001) Future climate change: implications for Indian summer monsoon and its variability. Curr Sci 81:1196–1207

Liu X, Chen B (2000) Climate warming in Tibetan Plateau during recent decades. Int J Climatol 20(4):1729–1742

Marinucci MR, Giorgi F (1992) A $2 \times CO_2$ climate change scenario over Europe generated using a limited area model nested in a general circulation model: 1. Present-day seasonal climate simulation. J Geophys Res 97:9989–10009

May W (2004) Potential future changes in the Indian summer monsoon due to greenhouse warming: analysis of mechanisms in a global timeslice experiment. Clim Dyn 22:389–414. doi:10.1007/s00382-003-0389-2

McBride J, Haylock MR, Nicholls N (2003) Relationships between the Maritime Continent heat source and the El Niño-Southern Oscillation phenomena. J Clim 16:2905–2914

Murari KK, Ghosh S, Patwardhan A, Daly E, Salvi K (2015) Intensification of future severe heat waves in India and their impact on heat stress and mortality. Reg Environ Chang 15:569–579. doi:10.1007/s10113-014-0660-6

Nakicenovic N, Swart R (2000) Special report on emission scenarios; IPCC working group-III special report. Cambridge University Press, Cambridge/New York. 599p

Nazrul Islam M, Uyeda H (2007) Use of TRMM in determining the climatic characteristics of rainfall over Bangladesh. Remote Sens Environ 108:264–276

Noguer M, Jones R, Hassell D, Hudson D, Wilson S, Jenkins G, Mitchell J (2002) Workbook on generating high resolution climate change scenarios using PRECIS. Hadley Centre for Climate Prediction and Research, Met Office, Bracknell, p 43

Oza M, Kishtawal CM (2015) Spatio-temporal changes in temperature over India. Curr Sci 109(6):1154–1158

Pal JS, Giorgi F, Bi X (2004) Consistency of recent European summer precipitation trends and extremes with future regional climate projections. Geophys Res Lett 31:L13202. doi:10.1029/2004GL019836

IPCC (2007) Summary for policymakers. Climate change 2007. Impacts, adaptation and vulnerability. In: Parry ML, Canziani OF, Palutikof JP, van der Linden PJ, Hanson CE (eds) Contribution of working group II to the fourth assessment report of the Intergovernmental Panel on Climate Change. Cambridge University Press, Cambridge, pp 7–22

Patwardhan SK, Krishna Kumar K, Kamala K, Preethi B, Revadekar JV, Rupa Kumar K (2008) Characteristics of Indian Summer monsoon in the warming scenario. In: Vinayachandran PN (ed) Understanding and forecasting of monsoons. Daya, Delhi, pp 150–157

Räisänen J (2003) CO_2- induced changes in atmospheric angular momentum in CMIP2 experiments. J Clim 16:132–143

Räisänen J, Hansson U, Ullerstig A, Döscher R, Graham LP, Jones C, Meier HEM, Samuelsson P, Willén U (2004) European climate in the late twenty-first century: regional simulations with two driving global models and two forcing scenarios. Clim Dyn 22:13–31

Rao S, Riahi K (2006) The role of non-CO_2 greenhouse gases in climate change mitigation: long-term scenarios for the 21st century. Multigas mitigation and climate policy. Energy J 3(Special issue):177–200

Revadekar JV, Kothawale DR, Patwardhan SK, Pant GB, Rupakumar K (2012) About the observed and future changes in temperature extremes over India. Nat Hazards 60(3):1133–1155

Revadekar JV, Hameed S, Collins D, Manton M, Sheikh M, Borgaonkar HP, Kothawale DR, Adnan M, Ahmed AU, Ashraf J, Baidya S, Islam N, Jayasinghearachchi D, Manzoor N, Premalal KHMS, Shreshta ML (2013) Impact of altitude and latitude on changes in temperature extremes over South Asia during 1971–2000. Int J Climatol 33:199–209. doi:10.1002/joc.3418

Riahi K, Gruebler A, Nakicenovic N (2007) Scenarios of long-term socio-economic and environmental development under climate stabilization. Technol Forecast Soc Chang 74(7):887–935. doi:10.1016/j.techfore.2006.05.026

Rowell DP (2005) A scenario of European climate change for the late 21st century: seasonal means and interannual variability. Clim Dyn 25:837–849

Rupa Kumar K, Ashrit RG (2001) Regional aspects of global climatic change simulation: validation and assessment of climate response over Indian monsoon region to transient increase of greenhouse gases and sulphate aerosols. Mausam 52:229–244

Rupa Kumar K, Krishna Kumar K, Ashrit RG, Patwardhan SK, Pant GB (2002) Climate change in India: observations and model projections. In: Shukla PR, Sharma SK, Ramana PV (eds) Climate change and India. Tata McGraw-Hill Ltd., New Delhi, pp 24–75

Rupa Kumar K, Krishna Kumar K, Prasanna V, Kamala K, Deshpande NR, Patwardhan SK, Pant GB (2003) Future climate scenarios. In: Shukla PR, Sharma SK, Ravindranath NH, Garg A, Bhattacharya S (eds) Climate change and India: vulnerability assessment and adaptation. Universities Press, Hyderabad, pp 69–127

Rupa Kumar K, Sahai AK, Krishna Kumar K, Patwardhan SK, Mishra PK, Revadekar JV, Kamala K, Pant GB (2006) High resolution climate change scenario for India for the 21st century. Curr Sci 90(3):334–345

Schär C, Vidale PL, Lüthi D, Frei C, Häberli C, Liniger MA, Appenzeller C (2004) The role of increasing temperature variability for European summer heat waves. Nature 427:332–336. doi:10.1038/nature02300

Semenov VA, Bengtsson L (2002) Secular trends in daily precipitation characteristics: greenhouse gas simulation with a coupled AOGCM. Clim Dyn 19:123–140

Shrestha AB, Wake CP, Mayawasky PA, Dibb JE (1999) Maximum temperature trends in the Himalaya and its vicinity: an analysis based on the temperature records from Nepal for the period 1971-94. J Clim 12(9):2775–2786

Shrestha AB, Bajracharya SR, Sharma AR, Duo C, Kulkarni A (2016) Observed trends and changes in daily temperature and precipitation extremes over the Koshi river basin 1975–2010. Int J Climatol:1–18. doi:10.1002/joc.4761

Sinha Ray KC, De US (2003) Climate change in India as evidenced from instrumental records. WMO Bull 52(1):53–59

Smith SJ, Wigley TML (2006) Multi-gas forcing stabilization with the MiniCAM. Energy J (Special issue #3):373-391

Stephenson DB, Douville H, Rupa Kumar K (2001) Searching for a fingerprint of global warming in the Asian summer monsoon. Mausam 52:213–220

Tebaldi C, Hayhoe K, Arblaster JM, Meehl GE (2006) Going to the extremes: an intercomparison of model-simulated historical and future changes in extreme events. Clim Chang 79:185–211

Turner A, Annamalai H (2012) Climate change and the south Asian monsoon. Nat Clim Chang 2:587–595. doi:10.1038/nclimate1495

Tyagi A, Goswami BN (2009) Assessment of climate change and adaptation in India. Clim Sense:68–70

Ueda H, Iwai A, Kuwako K, Hori ME (2006) Impact of anthropogenic forcing on the Asian summer monsoon as simulated by 8 GCMs. Geophys Res Lett 33. doi:10.1029/2005GL025336

Vidale PL, Lüthi D, Wegmann R, Schär C (2007) European climate variability in a heterogeneous multi-model ensemble. Clim Chang. doi:10.1007/s10584-006-9218-z

Voss R, May W, Roeckner E (2002) Enhanced resolution modeling study on anthropogenic climate change: changes in the extremes of the hydrological cycle. Int J Climatol 22:755–777

van Vuuren DP, Eickhout B, Lucas PL, den Elzen MGJ (2006) Long-term multi-gas scenarios to stabilise radiative forcing — exploring costs and benefits within an integrated assessment framework. Multigas mitigation and climate policy. Energy J 3(Special issue):201–234

van Vuuren D, den Elzen M, Lucas P, Eickhout B, Strengers B, van Ruijven B, Wonink S, van Houdt R (2007) Stabilizing greenhouse gas concentrations at low levels: an assessment of reduction strategies and costs. Clim Chang. doi:10.1007/s10584-006-9172-9

Wise MA, Calvin KV, Thomson AM, Clarke LE, Bond-Lamberty B, Sands RD, Smith SJ, Janetos AC, Edmonds JA (2009) Implications of Limiting CO_2 concentrations for land use and energy. Science 324:1183–1186

Xu YL, Huang XY, Zhang Y, Lin WT, Lin ED (2006) Statistical analyses of climate change scenarios over China in the 21st century. Adv Clim Change Res 2:50–53

Chapter 5
Climate Change: Central Himalayan Perspective

5.1 The Central Himalaya

The Central Himalaya (CH) located in the north-central part of the Indian subcontinent extending up to the Tibetan plateau is a region which assumes an important position in the climate map of the world and also displays high sensitivity to climate change. The rivers originating from the CH have access to perennial source of water

G.B. Pant et al., *Climate Change in the Himalayas*,
DOI 10.1007/978-3-319-61654-4_5

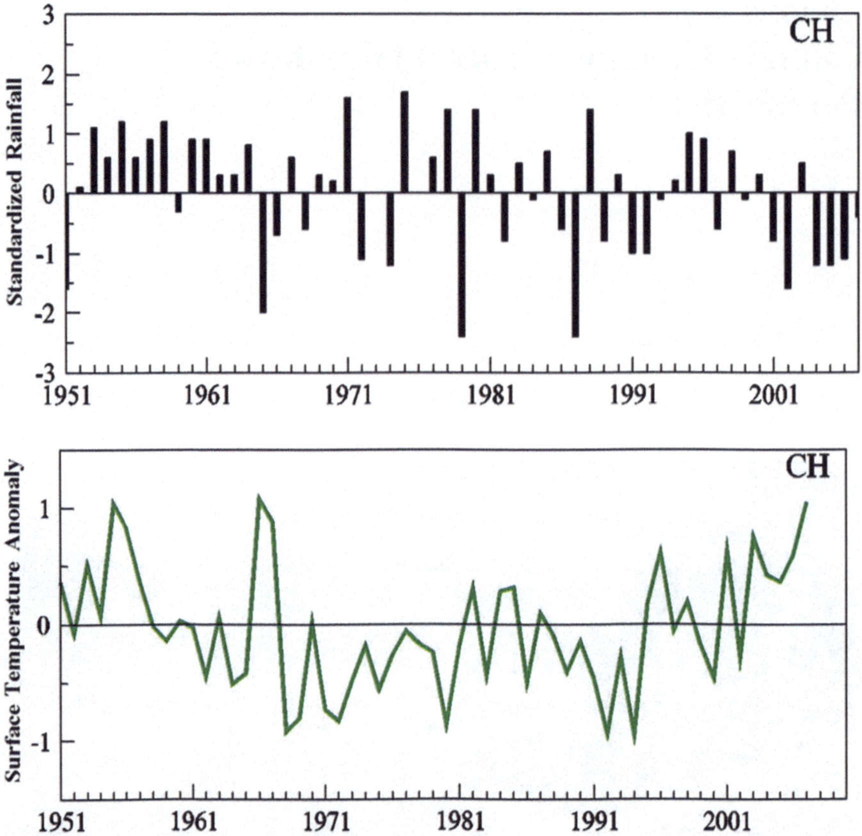

Fig. 5.1 Interannual variability of seasonal (JJAS) rainfall over central Himalaya (CH) for the period 1951–2007 (*top*) and annual average temperature (*bottom*) (Kulkarni et al. 2013)

from accumulated snow on the mountains and glaciers as well as the recharge of rivers and streams by seasonal rains and snowfall. In addition to abundant natural resources, the region possesses a highly esteemed natural beauty, cultural richness, and strategic location. The geology and geography of the region provides good scope for the assessment and evaluation of climate change impacts. In the subsequent discussions dealing with climate change impacts over the region the administrative boundaries of the state of Uttarakhand are assumed to be synonymous with the CH. For consistency of analysis, the peripheral stations such as Shimla in Himachal Pradesh along the western boundary of Uttarakhand are also included. The available model generated reanalyzed data sets describe the general climatology of the region and indicate an increasing trend in mean temperature. Figure 5.1 gives the seasonal rainfall and annual average surface temperature over central Himalayan region, for a period of 1951–2007. Starting from a brief introduction to the state of Uttarakhand in the context of Himalayan region, this chapter will deal

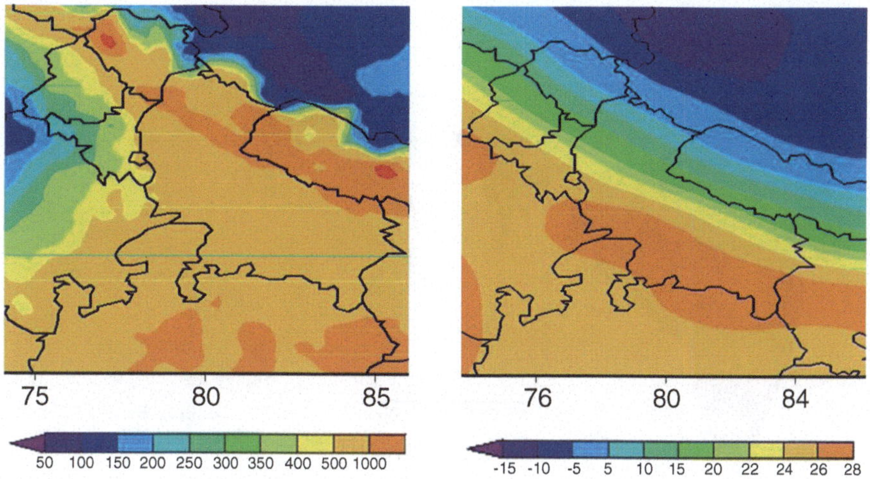

Fig. 5.2 Observed climatology of seasonal (JJAS) rainfall (*left*) and annual average temperature (*right*) based on 1961–1990 (Kulkarni et al. 2013)

with the current emission scenario and future potential for GHGs in the state. The observed climatology of monsoonal rainfall and the annual mean temperature for the period of 1961–1990 are given in Fig. 5.2.

5.2 The State of Uttarakhand

The state of Uttarakhand is located in the central parts of mighty Himalayan ranges covering about 86% of the area comprising mountain terrain and the 14% belonging to plain regions along the foothills with marshy land and humid climate (Bhabar and Tarai). The mountainous region extends from the Shiwalik hills in the southern periphery to the great Himalayan ranges in the north up to the Tibetan plateau with varying geological and ecological regimes. A map indicating the physical location of the state and an extended map of the state representing all districts is given in the Fig. 5.3.

The state has been carved out of the northern hill districts of CH and fertile lands of the state of Uttar Pradesh in the year 2000. It is endowed with a unique and highly fragile mountain ecosystem consisting of large variety of flora, fauna, and types of weather and climate. About 60% area of the state is under forest cover and is also a major source of freshwater for the subcontinent with two major river systems Ganga and Yamuna originating from the upper reaches of the state. The rivers originating from the Himalaya have access to permanent source of water from accumulated snow on the mountains and glaciers and also the seasonal recharge of rivers and streams by rain and snow.

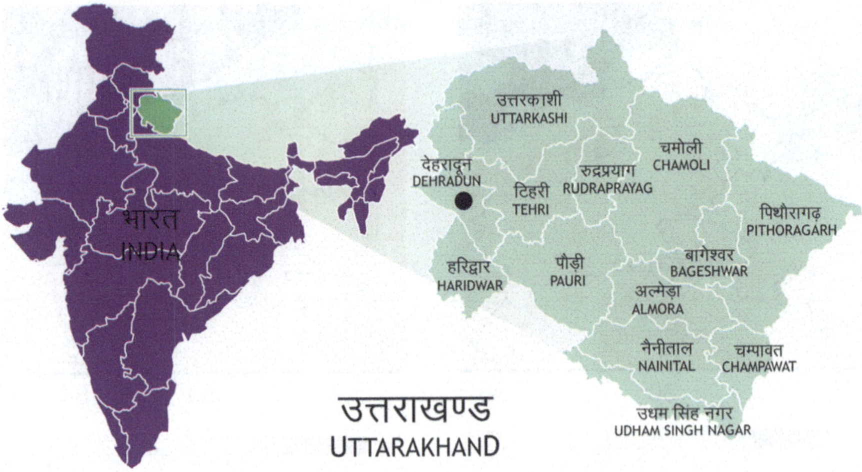

Fig. 5.3 Physical location and the administrative map of Uttarakhand

In addition to abundant natural resources, the state has highly esteemed natural beauty, cultural richness, and strategic location which provides ample opportunity for sustainable development while retaining and cherishing its clean environment and rich ecosystem. The climatic significance of the region has been recognized by the India Meteorological Department by designating it the status of a separate meteorological subdivision named as the "Hills of Uttar Pradesh" much before the state of Uttarakhand assumed its independent identity as a distinct hill state of CH.

5.3 Sources of Greenhouse Gases and Emission Scenario

It is now well recognized and accepted that the GHGs and other pollutants as drivers of climate change have been in continuous upswing since the recent industrial revolution identified by consumption-based developmental parameters on global as well as at country and smaller subdivisional scales. The role of GHGs in global climate dynamics and their manifestations on regional and local-scale climatic changes have resulted into mandatory monitoring, inventories, and mitigation strategies under numerous national and international efforts. A quick glance over major sectors of society contributing to total anthropogenic emission of GHGs as of the year 2004 in terms of the CO_2 equivalent reported by the IPCC (2001) is given in the following Table 5.1.

Periodic update and national inventory of sources and sinks of major GHGs using comparable methodologies as agreed upon by the COP is a mandatory requirement of the UNFCCC. All national governments that are parties to the UNFCCC

Table 5.1 Global estimates of contributions of major activity sectors to the GHG emissions

Sectors	Percent of total
Energy generation and supply	25.9
Industry	19.4
Forestry	17.4
Agriculture	13.5
Residential and commercial building	7.9
Solid waste and wastewater	2.8
Others	13.1

and/or the Kyoto Protocol are required to submit annual inventories of all anthropogenic GHGs from sources and removal from the sinks (http://www.worldwatch.org/node/6449). Scientifically conducted national inventories are essential to update and refine the estimates of global emission statistics for climate change study. They also help in monitoring the quality and reliability of impact assessments, as well as provide a baseline for individual nations to develop their future emission trajectories. These measurements may find a significant place in identifying and evaluating mitigation strategies for GHG emissions and also act as input to model simulations of future climate change scenarios on regional scale.

In India, the GHGs and other components of air pollution are being monitored by various government agencies at different levels. These updated inventories are released periodically by the Ministry of Environment Forests and Climate change, Government of India (http://www.envfor.nic.in) being a nodal agency. These inventories provide basic inputs at national level and act as core data source for many regional environmental planning mechanisms along with their significance as authentic inputs to various international negotiations. Though the data on these significant parameters for CH is yet to be generated with acceptable quality and length of time, the preliminary records maintained by state authorities indicate that at current level of activity the region does not appear as a hot spot of GHG emissions in any of the above-mentioned sectors.

In the meantime, while the data collection network for major emitters of GHGs are being established in the state, creation of a data warehouse on other relevant parameters and related information is very essential. In addition to the observations of surface meteorological data collected by the IMD, rainfall, and temperature at many places is also being routinely collected by the Central Water Commission (CWC), the Indian Council of Agricultural Research (ICAR), and the Defense Research and Development Organization (DRDO) at the locations of their interest particularly at higher altitudes. The CWC have used extensive data over Ganga–Yamuna catchment to provide basic data for flood forecasting and also to calculate the Probable Maximum Precipitation (PMP) and Recurrence Periods for floods within the catchment area of major hydropower projects. In view of many developmental programs being vigorously implemented over the region, it is worthwhile to look at the status of a few important GHG emitting sources in the state. In the following paragraphs, a brief summary of five main activities with high GHG emission potential is being presented in a subjective manner due to the nonavailability of authentic long period data and analysis.

5.4 Major Sectors with Emission Potential

5.4.1 Industry

Most part of CH consists of the mountainous region having practically no major industry with substantial polluting or GHGs emission potential. At the time of creation of a new state of Uttarakhand, there have been no major industrial activity except for a few scattered locations in the plain areas mostly confined to sugar mills, paper pulp, and medium-sized agro-based industries. With new wave of industrial development, many industries including some with high GHG emission potential have cropped up in the notified development zones particularly in the areas around Haridwar, Roorkee, Dehradun, and Udham Singh Nagar. Industries are essential for an overall development of the state mainly to provide income and employment as well as to fulfill the aspirations of the local people and also to add to gross national productivity. Requirement for an environment-friendly industrial policy catering to the needs of eco-friendly developmental norms is a must for the region in view of its high vulnerability to environmental changes at all spatial scales. Limited data and current understanding of climate change-induced environmental impacts is not sufficient to provide reliable scientific tools to assess the vulnerability and impending damage levels.

5.4.2 Power Generation

As a source region of mighty rivers Ganga, Yamuna, and their tributaries, the region of CH has a unique distinction of possessing maximum potential for power generation through hydropower. Tehri and Nanak Sagar hydropower projects are the major sources of hydropower in the state. In addition, there are many other small and micro-hydropower projects in the process of development and commissioning. With abundance of water resource, favorable mountain slopes and relatively easier access to the project sites, the hydroelectricity could be perhaps the most convenient and economically viable green energy source subject to the condition that the environmental and safety issues are taken care of. Hydropower helps to reduce the GHG emission since it provides a non-emitting substitute to fossil fuel and also helps the environment by sequestering carbon in water reservoirs as the forests do over the land thus acts as carbon sink. There are also counter arguments that the major dams contribute to global warming by way of the release of CH_4 gas from the stored water and marshy lands created in the periphery of the water reservoirs feeding the power plants.

There also exist good potential for solar, geothermal, wind and bioenergy generation at scientifically explored and thoroughly studied locations. The contribution of renewable green energy other than the hydropower within the region at present is almost negligible. An analysis of the data from the Solar Energy Atlas of India

(Mani and Rangarajan 1982) suggests that the solar power for the generation of electricity to feed the power grid may not be economical or feasible over the region. This assumption is based on the fact that the annual solar power potential over the region is relatively small and the cost of coupling to the grid from scattered locations and rough terrains will be very high. In the meantime, there exists very high potential of solar energy applications at local level specifically, for day-to-day domestic needs. Dust-free environment and intense daytime solar radiation on clear sky days can help in better maintenance of rooftop solar water heaters and battery charged house lanterns and open area street lights.

Wind energy resource assessment for the country done by the National Institute of Wind Energy, Chennai has suggested viable wind energy potential mostly confined to the coastal regions of the Indian peninsula and the Western Ghat mountain ranges and some parts of Aravalli ranges (http://niwe.res.in/). The mountain regions of CH are mostly the regions of weak and variable winds as far as the large scale prevailing wind profile is concerned. However, there are many hilltops on the windward side of the mountain ranges, where wind energy can be tapped for local needs. The basic requirement of a potential wind energy site is that it should have persistent wind speed of more than a critical threshold value during the year and preferably a low variability of wind direction. As an example, the maximum surface wind profile of windward side location at the ARIES, Nainital presented in Fig. 5.4 shows a reasonably good wind power potential for local tapping.

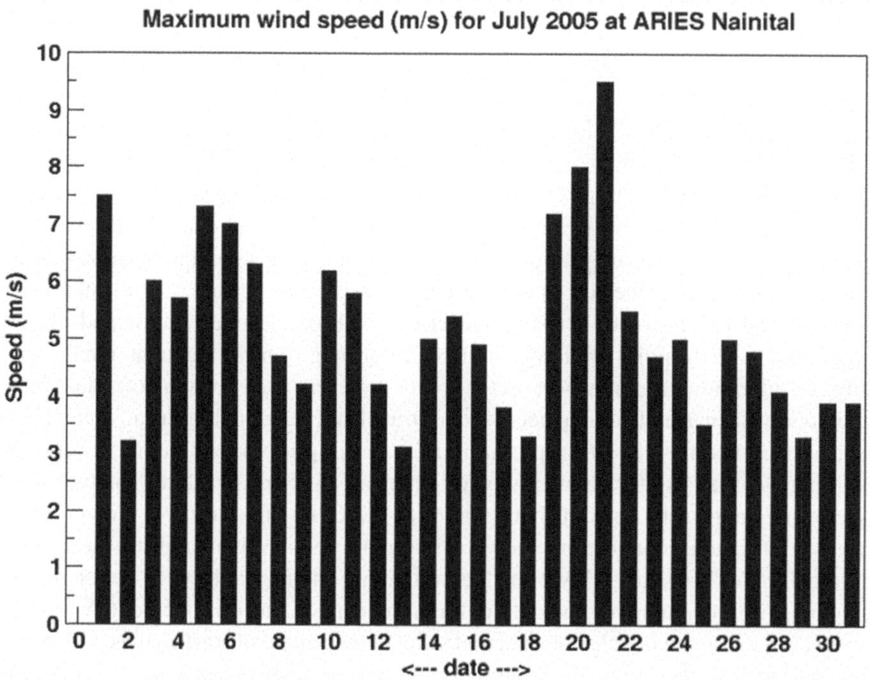

Fig. 5.4 Profile of daily maximum wind speed (ms^{-1}) for July 2005 at Manora Peak, Nainital

Preliminary survey and geophysical analysis suggests that some regions of the Uttarakhand state, particularly the locations in the districts of Uttarkashi, Chamoli, and Pithoragarh are promising for the exploitation of the geothermal energy (Sharma et al. 1996; Bharadwaj and Tewari 2008; Dimri 2013). In the neighboring state of Himachal Pradesh, there are known hot water springs at many sites in Manikaran area of Kulu district and some other nearby locations. At present, the exploration for geothermal energy sources with new sites and potential including the efforts for tapping the available energy at the known sites is negligible for CH. In case of bioenergy, the basic potential is high in the villages and towns with their local biodegradable waste products which can be harnessed using efficient and appropriate technology. The bioenergy technology is simpler to install but continuous supply of appropriate inputs and proper maintenance particularly over village environment may pose problems. Additionally, the importance of bioenergy as a solid waste management alternative is very relevant and needs to be pursued vigorously.

Overall, the region has high renewable and nonpolluting energy sources to meet the increasing energy demands posed by recent developmental activities. It is also necessary that the energy requirements of the future must be framed keeping in mind the likely additional needs arising out of the changing climate while stabilizing the contributions of the region to total GHG emissions. In Uttarakhand, the state Renewable Energy Development Agency (UREDA) (http://www.ureda.uk.gov.in) can play a proactive role in popularizing the renewable energy options with encouragement and support from state and central governments as well as public and private sector enterprises.

5.4.3 *Transportation*

The CH regions by and large consist of rural habitats and villages with eco-friendly settlements. With the formation of new State and the emergence of new wave of development, the state is in a rapid pace of advancement towards urbanization in plain areas and also the towns spread across large hilly terrains. With increased demand and fast growth for road connectivity to high altitude regions and places important from the point of view of tourism, pilgrimage, and strategic national security considerations, the transport sector possesses a very high growth potential. With the present scenario of limited access to air travel and restricted rail road connectivity, only choice for mass transportation is to develop motorized vehicle transport within cities and also for connectivity to remote areas across the state. The transport sector may appear to be contributing a lot to the GHG emission when one comes across traffic jams, congestion, and vehicular pollution in urban areas resulting into microclimatic hot spots. However, the situation is manageable with better traffic planning, improvement of road quality for uninterrupted traffic movement. Alternate sources of energy using cleaner fuel and strict enforcement of traffic rules and existing laws may further improve the situation. There are many possible steps necessary to minimize GHGs emission from the transport sector and improve the urban air quality with a scope to implement them in phased manner.

5.4.4 Agriculture

The agriculture in central Himalayan states is distinctly different for the plains and hilly regions. In hilly areas, large percentage of population solely depends on agriculture with a small fraction of about 10–15% land area being available for cultivation. Agriculture in hills is mostly rainfed and organic and traditional in nature with little use of fertilizers, pesticides, and mechanization. The agricultural produce being mostly of native varieties of food grains with relatively smaller crop yield. Therefore, the agricultural practices in the hills contribute a small fraction to the total GHG emission of the state. In the plain regions of Himalayan foothills or the valley sites in the mountain ranges, agriculture is mostly under irrigation of some kind or the other from the sources such as river/lake water, groundwater, ponds, canals, and dams and most of the areas also use modern practices.

In the recent years, many areas are adopting intensive agricultural practices using modern technology with emphasis on cash crops such as sugarcane and high yielding varieties of wheat, rice, cotton, fruits, vegetables, and flowers. Most of the agricultural land in the plain regions is reclaimed from marshy patches and marginal forests with abundance of natural methane and CO_2 emission prior to land reclamation. Therefore, the emission of methane from rice fields or the carbon dioxide emission from agricultural practices might not have increased substantially from their natural emission levels as a result of adoption of intensive agriculture and modern land use practices in the recent decades.

This situation may however change once the intensive agricultural practices with irrigation and fertilizer use become the norms to meet the increased food demands and better income for farmers. Burning of agricultural residue may also add substantially to CO_2 emission, if proper care is not exercised by the farming community to maximize its use as agricultural manure. Another important point to remember in context of agriculture and climate change is the loss of area under forest to cultivated land. As it is well known that the area under perennial forest cover provides a major sink of CO_2 in the atmosphere, deforestation for agriculture must appropriately substitute for the green cover by afforestation and agro-forestry.

5.4.5 Firewood and Biomass Burning

Many studies using the data of selected pilot studies extrapolated over a wider region have pointed out that the burning of firewood, cow dung, and other local biomass for domestic use in villages and small towns may have considerable CO_2 emission potential. As far as the CH with relatively smaller population density is concerned, the domestic use of firewood and biomass burning is not likely to have significant contribution to the GHGs. There is no quantitative data available to assess the current status and pattern of change in the domestic emission of GHGs. Strict laws relating to cutting of trees and provision of restricting the use of forest products for domestic needs along with adequate availability of cooking gas and electricity for household applications is visibly altering the scenario in the recent years.

There is ample scope to collect data and monitor the production of GHGs, particularly the emissions as a result of forest fires and the natural emissions from the forest ecosystem. The practice of firewood burning by the people particularly the villages in remote hilly areas during winters to keep their houses warm can be managed with the provision of rural electrification. If the green energy and conservation policies are adopted and followed vigorously, it is possible that most rural areas in the state may attain low carbon emission status in coming decades with decreasing dependence on CO_2 emitting energy sources for domestic requirements. To control the GHG emission in urban areas, modern scientific methods and proper planning are required for management of solid waste and dry biomass burning in open need to be completely avoided.

References

Bharadwaj KN, Tewari SC (2008) Geothermal energy resource utilization; perspective of the Uttarakhand Himalaya. Curr Sci 95:7

Dimri VP (2013) Geothermal energy resources in Uttarakhand, India. J Ind Geophys Union 17(4):403–408

IPCC (2001) Climate change 2001: the scientific basis. In: Houghton JT et al (eds) Contribution of working group I to the third assessment report of the Intergovernmental Panel on Climate Change. Cambridge University Press, Cambridge/New York. 881p

Kulkarni A, Patwardhan S, Krishna Kumar K, Ashok K, Krishnan R (2013) Projected climate change in the Hindu Kush–Himalayan region by using the high-resolution regional climate model PRECIS. Mt Res Dev 33(2):142–151

Mani A, Rangarajan S (1982) Solar radiation over India. Allied Publishers, New Delhi, p 646

Sharma SC, Shukla SN, Dhaulakhundi AB (1996) Geothermal water resources of Tapoban fields, Chamoli District of U.P. and its utilization. Geol Surv India Spec Publ 21:149–154

Chapter 6
Central Himalaya: Climate Change Signatures

6.1 Meteorological Observations for Climate Change Study

There are only a few studies on the status of climate change over the CH region based on a spatially homogeneous long series of meteorological data. Paucity of long and continuous records of weather data, primarily due to inaccessibility to remote and high altitude locations is the main reason for the absence of high resolution climatology of the region. The major meteorological factors which form the basis of regional climatology are dominated by summer monsoons particularly on the southern hillslopes. During winters, the western disturbances are the primary factor and equally important are the orographic and convective thunderstorms that occur in the afternoon during pre- and post-monsoon seasons. The influence of variability in mean temperature and rainfall conditions along the ridges and valleys, windward and leeward side of mountains, and their slope and aspects are clearly detectable in the local meteorological data. In general, there is an average decreasing trend from south to north in temperature, the gradient being sharpest when the plain regions start rising to varying altitudes across the mountains. This trend is substantially modified by the seasonal migration of weather systems.

Only studies conducted so far using long series of data for the state (Pant and Borgaonkar 1984; Pant et al. 1999; Borgaonkar et al. 2011) suggest that the annual rainfall of stations across the Uttarakhand state displays a high degree of spatial coherence. This inference points towards spatial homogeneity in interannual

G.B. Pant et al., *Climate Change in the Himalayas*,
DOI 10.1007/978-3-319-61654-4_6

variability of rainfall amounts across the state indicating that the rains from large-scale systems like monsoon and western disturbances dominate the spatial domain of rainfall pattern in the state. The composite rainfall time series for these stations does not show any long-term trend or periodicity. A relatively dry period of below normal rainfall during 1925–1975 as seen in the data suggests an episodic nature of the state level rainfall amounts, similar to the pattern seen in the all India monsoon rainfall series given by Fig. 1.6 in Chap. 1.

An examination of temperature data of Mukteshwar, a high altitude mountain station (2346 m amsl) with high quality continuous record since 1901 displays marked increase in winter maximum and mean temperature since 1935. This increase in mean temperature is in broad agreement with general increasing trend in the all-India temperatures and also the widely known signal of global warming; though, a part of this increase in the recent years may be attributable to local activities such as deforestation and change in the observatory site with inadequate location correction. Figure 6.1 presents the long-term maximum temperature trend over Mukteshwar for the period 1901–2012 during the winter (DJF) season as an example (data courtesy: IMD).

In a recent study, Borgaonkar et al. (2011) updated the data used by earlier authors (Pant et al. 1999) for eight stations of Uttarakhand to construct the Western

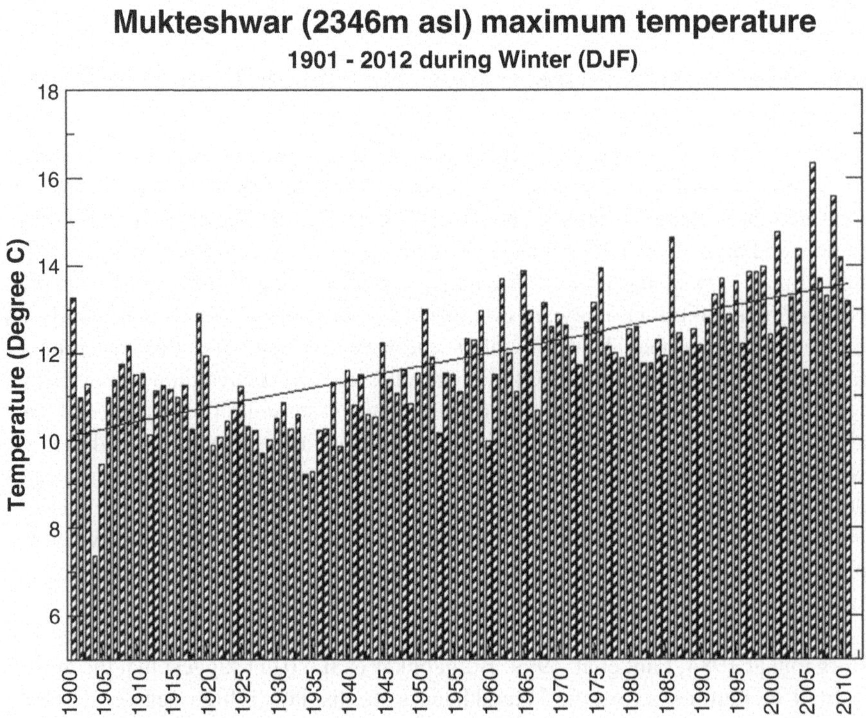

Fig. 6.1 Mukteshwar maximum temperature for winter season for a period of 1901–2012 shown by vertical bars. *Solid line* depicts linear trend

Himalayan composite rainfall series (1901–2005). The stations in Uttarakhand are Mussoorie, Mukteshwar, Dehradun, Nainital, Pauri, Pithorgarh, Joshimath, and Almora. Two important stations with continuous reliable data from the west of Uttarakhand are also included. These are Shimla (Himachal Pradesh) and Srinagar (Jammu and Kashmir). Temperature trends are studied using mean monthly surface air temperature data for three stations in Western Himalaya, namely, Shimla, Mussoorie, and Mukteshwar for the period 1901–2003. All these stations are high altitude stations with heights of more than 2000 m amsl, and are latitudinally distributed across the state with Mukteshwar being the lowest and Shimla the highest latitude station from south to north.

The rainfall and temperature data of these stations show high degree of spatial coherence, as inferred from statistically significant intercorrelation in the rainfall data sets of annual as well as the four standard seasons. Composite temperature anomaly series shown below for annual, winter (DJF), pre-monsoon (MAM), Monsoon (JJAS), and Post-monsoon (ON) seasons clearly indicates a rising trend in mean temperature during the four seasons and also in annual value. They have also used the monthly rainfall values of nine stations, eight from Uttarakhand and one station Shimla from the neighboring state of Himachal Pradesh and constructed a composite rainfall series for the period 1871–2000. This study with best possible data coverage corroborates the earlier findings that the intercorrelation of rainfall among the stations are positive and highly significant for different seasons. They further conclude that the trend analysis for different seasons clearly shows that none of the seasons display significant trend in rainfall.

The mean climatology of two representative hill stations, namely, Mussoorie and Mukteshwar are presented in Fig. 6.2. In terms of rainfall, the contribution of monsoon rain to the annual total rainfall is much higher than the winter and other transition seasons for both the stations. Mukteshwar is south-east of Mussoorie; therefore, the monsoon season is generally active there slightly earlier than Mussoorie with active monsoon in July and August. The temperatures, particularly the maximum temperature, shows an annual cycle with prominent influence of day time cooling due to cloudiness and rain during monsoon season.

Annual highest values of daily maximum and minimum temperatures are analyzed for a period of about more than 30 years, for the three stations Srinagar, Shimla, and Mukteshwar as given by Fig. 6.3. It can be observed from the figure that a gradual but increasing trend is seen in the maximum as well as minimum temperatures except a few anomalies.

Seasonal mean surface temperature series over the Western Himalayan region for the total period 1901–2007, also shows an increasing trend. The recent four decades since 1961 have also been analyzed for mean surface temperatures trends separately, and it is found that the recent five decades are demonstrating faster increase in the mean surface temperature as compared to the previous six decades as represented by Fig. 6.4. Larger temperature anomalies are also seen in the recent five decades which is an indicator of boosted atmospheric dynamics that controls various physical and chemical processes in the atmosphere. Correlation estimated between the stations Mukteshwar and Mussoorie is given in Table 6.1 for different seasons. Mean temperatures are found to be highly correlated for winter season.

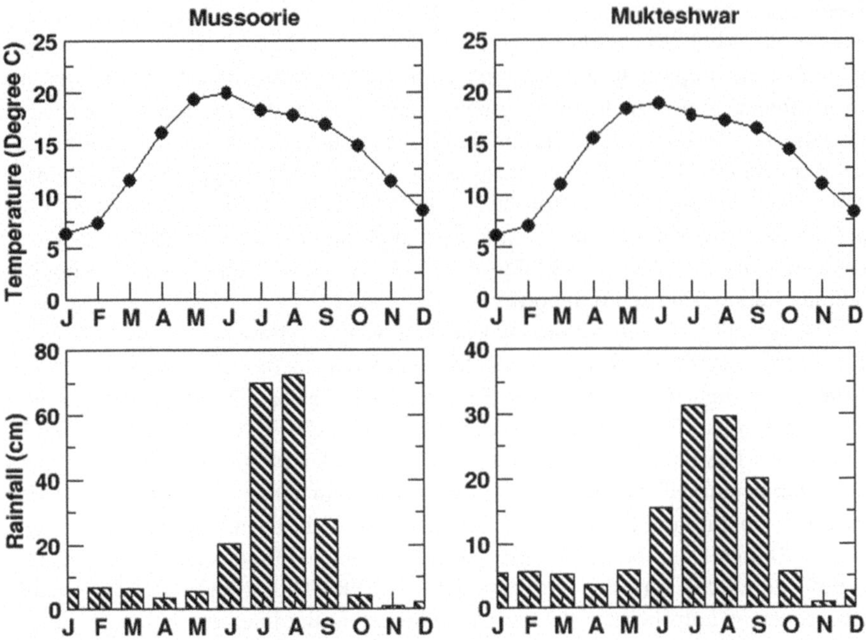

Fig. 6.2 Average monthly variation of mean surface temperature and rainfall of Mussoorie (*left column*) and Mukteshwar (*right column*) based on the IMD data period 1901–2003 and 1871–2000, respectively

It is interesting to note that the stations Mussoorie and Mukteshwar have highly correlated temperatures during winter suggesting that the large-scale western disturbances are the primary source of winter weather in terms of rain/snow with similar influence at both stations. Highest warming trend is seen in winters followed by the post-monsoon season. The warming trend during monsoon season is almost undetectable perhaps due to the complex relation between large-scale monsoon currents and the local temperatures.

6.2 Past Climates Over CH: As Derived from the Growth Rings of Trees

Extensive geographical extent of Himalaya from west to east and sharp altitude gradients from north to south gives rise to differences in temperature and precipitation regimes. These differences produce distinct influence on certain long-lived biospheric and hydrospheric processes operating locally as well as related to global influences. These influences are detectable in the floral composition, tree growth parameters, flowering pattern, position of the upper tree line, glacier margins, and characteristics of snowmelt discharge into rivers. It is most likely that some of the

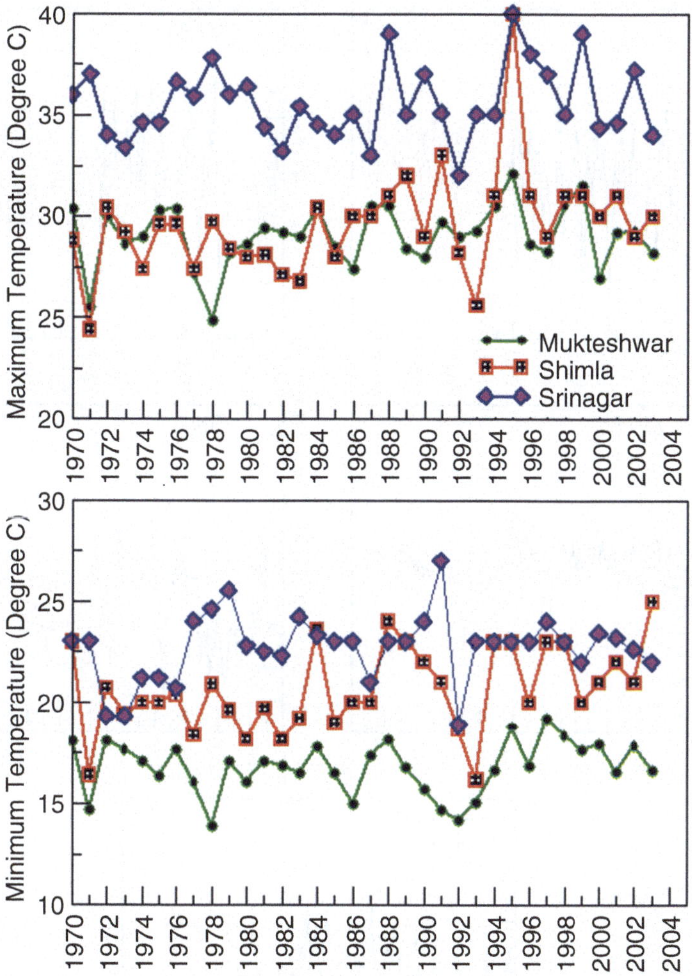

Fig. 6.3 Annual highest values of Maximum and Minimum temperature using daily IMD data at Mukteshwar, Shimla, and Srinagar observatories

changes occurring in climate on global and regional scales over a period of time may have left their permanent footprints on some of the biospheric natural indicators. These indicators may provide indirect information for a much longer period than the observed meteorological data which is essential to study the large-scale climate changes over the region in longer time frames beyond the realm of meteorological observations.

Time series on stable isotope ratios (C^{13}/C^{12}, O^{18}/O^{16}, D/H) from water or cellulose or tree sap is one of the potential sources of valuable information; however, hardly any systematic data is available for the Indian region on these parameters. Observed meteorological data on precipitation and temperature are generally available only for about a century or shorter duration. This is particularly true for most of

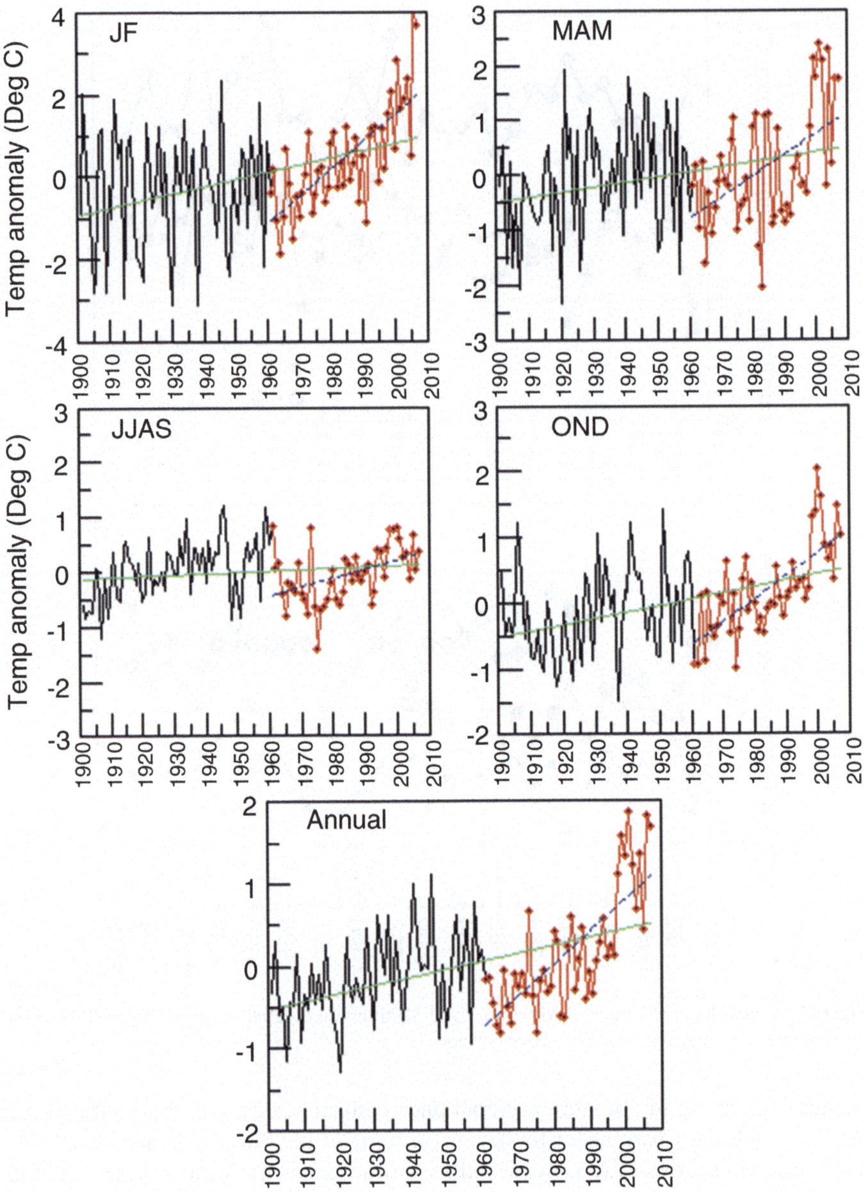

Fig. 6.4 Mean surface temperature series over Western Himalaya for different seasons with linear trends for the total period 1901–2007 (*solid line*) and recent five decades 1961–2007 (*dashed line*) based on IITM data (freely available at www.tropmet.res.in)

Table 6.1 Trend (°C/100 years) in mean temperature over Uttarakhand stations

Station	DJF	MAM	JJAS	ON	Annual
Mussoorie	1.4	0.4	−0.1	1.1	0.5
Mukteshwar	1.4	0.3	0.0	0.9	0.6
Correlation between the station	0.91	0.88	0.78	0.78	0.79

the observatory stations in the state of Uttarakhand. In view of this, data from selected climate-sensitive natural systems (trees, glaciers, soil, pollens, etc.) can be derived and used to reconstruct the seasonal temperature and rainfall estimates of the past beyond available meteorological observations (Pant 1979). The branch of science using growth rings of trees immediately prior to available meteorological data as proxy, for the reconstruction of the climate of the past is called, Dendroclimatology. Growth pattern of specific tree species with well-marked annual growth rings from the old living coniferous trees such as fur, spruce, pine, deodar, and juniper from Himalayan region are used to decipher rainfall and temperature data for long periods of time spanning over the entire age of a long-living tree. Available short period rainfall and temperature data series from a nearby station with reliable continuous record are used to establish statistical relationship between tree growth and meteorological parameters. These relations are regressed upon the past tree ring data to reconstruct rainfall and temperatures values of the past.

Using a total of five tree ring width chronologies of Himalayan conifers of Deodar (Cedrus deodara) and spruce (Picea smithiana), Borgaonkar et al. (2011) created a master tree ring chronology of Western Himalaya after averaging all the five series. In this chapter, the master chronology superimposed over smooth line is cubic spline filter with the 30 years wavelength. It indicates that higher tree growth during last few decades is attributed to the warming trend over the region. However, this chronology also indicates low growth periods during Little Ice Age (AD 1450–1850). Thus, the ring width index is reliable indicator of temperature trends (Borgaonkar et al. 2011; Singh and Yadava 2000). The data on ring width index for about a last century show the patterns comparable to the temperature patters of Mukteshwar and Mussoorie stations.

A growth surge observed in coniferous trees from central Himalaya (sites in Uttarakhand and Himachal Pradesh) since 1940 as indicated by anomalously high ring width index value is a strong indicator of an association between tree growth surge in Himalayan conifers and increasing temperatures which have perhaps provided more conducive growth environment to the tree species examined for the region. Singh and Yadava (2000) have also demonstrated a significant increase in the growth of pine trees since 1950 from Chirbasa near Gangotri in Uttarakhand in relation to recent atmospheric warming trends. In addition to this, there are many studies in literature suggesting warming trends in Tibetan plateau and Nepal Himalaya with corresponding tree growth pattern from those regions. In view of the fact that the precipitation during monsoon season over the country shows an indication of decadal scale dry and wet anomalies as discussed in Chap. 1, it is desirable that the information from all proxy sources is consolidated to detect the precipitation behavior in the central Himalaya (Uttarakhand) during the little Ice Age and the Medieval Warm Period to understand the mechanism.

Some interesting recent studies (Kotlia et al. 2014, 2016a, b) has used speleothem record from a cave site in central Himalaya to profile the precipitation variability for the last 4000 years. These studies highlight the interplay between the northward moving Inter Tropical Convergence Zone (ITCZ) and the strength and extent of westerly wind regimes. They suggest that since most of the winter precipitation in central Himalaya is due to the weather systems of westerly origin and the monsoon precipitation with its origin in the warm southern oceans; therefore, the proxy climate signals from this region may help understand the tropical and middle latitude interaction and delineate the influences on Himalayan climate. At present, there are very limited studies over the region to quantitatively reconstruct the climate variables for few thousand years and draw definite conclusions. Therefore, there is a need to correlate the decadal scale variability from tree rings, speleothems, etc. with historical records, archeological evidences, and lake sediment profiles.

References

Borgaonkar HP, Sikder AB, Ram S (2011) High altitude forest sensitivity to recent warming: a tree-ring analysis of conifers from Western Himalaya, India. Quat Int 236:158–166

Kotlia BS, Singh AK, Joshi LM, Dhaila BS (2014) Precipitation variability in the Indian Central Himalaya during last ca. 4,000 years inferred from a speleothem record: Impact of Indian Summer Monsoon (ISM) and Westerlies. Quat Int. doi:10.1016/j.quaint.2014.10.066

Kotlia BS, Singh AK, Sanwal J, Raza W, Ahmad SM et al (2016a) Stalagmite inferred high resolution climatic changes through pleistocene-holocene transition in Northwest Indian Himalaya. J Earth Sci Clim Change 7:338. doi:10.4172/2157-7617.1000338

Kotlia BS, Singh AK, Zhao J-X, Duan W, Tan M, Sharma AK, Raza W (2016b) Stalagmite based high resolution precipitation variability for past four centuries in the Indian Central Himalaya: Chulerasim cave re-visited and data re-interpretation. Quat Int. doi:10.1016/j.quaint.2016.04.007

Pant GB (1979) Role of tree-ring analysis and related studies in Palaeoclimatology: preliminary survey and scope for Indian region. Mausam 30:439

Pant GB, Borgaonkar HP (1984) Climate of the hill region of Uttar Pradesh. Himalayan Res Dev 3(1):13–20

Pant GB, Rupa Kumar K, Borgaonkar HP (1999) Climate and its long term variability over the western Himalaya during the past two centuries. In: Dash SK, Bahadur J (eds) The Himalayan environment. New Age Intl. Publ., New Delhi, pp 171–184

Singh J, Yadava RR (2000) Tree-ring indications of recent glacier fluctuations in Gangotri, Western Himalaya, India. Curr Sci 79:1598–1601

Chapter 7
Climate Change Impacts: Central Himalaya

7.1 The Glaciers

Himalayas are the high altitude areas of perennial snow accumulation possibly the fragmented remnants of quaternary ice-age era over the earth. The seasonal loss of snow due to melting is periodically replenished by changing weather systems which provide fresh snow practically every year with varying degree of thickness and duration. The outward and downward moving ice due to internal dynamics and thermodynamics of a complex system produces rivers of ice, known as the Glaciers.

G.B. Pant et al., *Climate Change in the Himalayas*,
DOI 10.1007/978-3-319-61654-4_7

These are confined to narrow spaces or valleys spread across the length of the mountain ranges. The total area of ice mass and depth of accumulation varies significantly in time and space with a possibility that a persistent change may occur for an extended period of time. These changes may appear in the form of advancing or receding glacier margins which can be quantitatively demarcated by in situ measurements and remotely sensed aerial surveys and satellite observations. Himalaya being the longest east-west oriented mountain range of the world has the largest snow mass outside the polar region and is estimated to contain about 10,000 glaciers along its southern slopes across its entire length. Himalayas are more intensely glaciated mountains than the higher latitude glaciers in Alps and Rockies mainly due to the abundance of higher altitude mountain ranges and also a continuous moisture supply from the prevailing weather systems.

The glaciers are inhomogeneously distributed along the length of Himalaya from west to east and concentration and spread of glaciers is much higher in the Western Himalaya (WH) than the eastern part. The Himalayan glaciers have been the subject of exploration, documentation, and scientific study by glaciologists, geologists, hydrologists, and survey personnel for a long time; however, due to enormity of the problem even the inventory of the entire Himalayan glaciers is not yet complete. The Survey of India, Geological Survey of India, Wadia Institute of Himalayan Geology, Defense Research and Development Organization, and the Indian Space Research Organization have regular programs of glacier inventory, monitoring, and research studies. In addition, few universities and academic institutions periodically undertake specific scientific study on selected glaciers under focused scientific projects. Considering the importance of the subject and requirement of major resource planning and management infrastructure, an integrated study of Indian glaciers particularly over WH deserves to be a top priority research area for the country.

7.1.1 Role of Cryosphere in Climate Change

The most important impact of climate change on the cryosphere are the melting of polar ice caps, changes in glacier ice mass balance, erosion of glacier margins, and retreat of glaciers in many parts of the world. Increasing ocean water temperature and consequent increase in total water volume and large flow of polar cold water to tropical oceans due to melting of polar ice caps may contribute to the increase in mean sea level. This influence may be further enhanced due to the albedo effect explained earlier. Addition of cold polar ice melt water to tropical warm oceans may also alter the flow pattern of deep ocean currents and disturb the ocean conveyor belts. Resultant perturbations of tropical ocean dynamics may have repercussions on large-scale ocean-atmosphere circulations such as El-Nino and monsoons which may have significant influence on the future climates.

The data from Antarctica, North Polar Regions, and mountains of Europe and North American continent support an enhanced loss of continental and polar oceanic ice volume in the recent past which is considered as an example of direct impact of global warming on the cryosphere. Precise data on the magnitude and sign of glacier margin movements over the Himalaya is limited to very few glaciers. Analysis of satellite imagery which is being increasingly used for this purpose can only provide a synoptic view over a wider area. For example, an analysis of satellite imagery by Ageta and Kadota (1992) suggests that almost about 67% of out of the total number of glaciers studied in the Himalaya have retreated in recent years which will have to be verified against ground truth over a target region. In some parts of Nepal Himalaya, the process of retreating glaciers is found to be faster and at an estimated rate of about 10 m/year.

The IPCC in its 2007 assessment report have stated that the shrinking of glaciers is one of the major concerns arising out of climate change. The report emphasizes that the mountain glaciers have an alarming rate of shrinking in area and substantial reduction in their snow volume. The projection of sharp decrease in Himalayan glaciers and their faster rate of retreat in recent decades have created obvious scientific controversy and national concern since snow and ice in Himalaya is very crucial for freshwater supply to one of the world's largest populated region. Recently released IPCC WG-1 2013 Report states that "since the last report (AR4-2007) almost all glaciers worldwide have continued to shrink as revealed by the time series of measured changes in glacier length, area, volume and mass. This includes the glaciers of Asian mountains." They further state that "the current glacier extents are out of balance with current climatic conditions, indicating that glaciers will continue to shrink in the future even without further temperature increase."

7.1.2 *Retreating Glaciers of Central Himalaya: Some Observations*

In the study of glaciers, the data of retreat or advance of glaciers, mass balance, ice volume, snow depth, and other related meteorological parameters are required to be routinely measured and documented. Considering the logistic difficulties and an enormously large area of the region, the current level of available data is in fact a microscopic fraction of the complete information. Some of the important glaciers of Central Himalaya (CH) have been independently studied by various groups and the data of earlier periods have many discontinuities in space and time. These parameters measured and reported by different groups over a period of time depict large to marginal variability in the cases studied. To assess the current status of Himalayan Glaciers, the data from many published papers and reports is examined specifically to find out the magnitude of retreat noticed in the glaciers of CH. Recently, a study group appointed by the Office of the Principal Scientific Advisor to the Government of India (RSG-GOI-2011) have summarized the glacier inventory and prepared a status report on the studies of Himalayan glaciers.

A critical analysis based on the limited data on CH glaciers suggests a recede in glacier margin to the extent of about 17 m/year (Table 7.1). The report also highlights that most of the glaciers have been passing through a phase of recession though there

Table 7.1 Retreat data on major glaciers of CH

Name of glacier	Period (years)	Average retreat (m/year)
Bandar punch	1960–1999 (39)	25.5
Jaundar Bamak	1960–1999 (39)	37.3
Jhajju Bamak	1960–1999 (39)	27.6
Tilku	1960–1999 (39)	21.9
Gangotri	1935–1999 (64)	19.0[a]
Bhrigupanth	1962-1995 (33)	16.5
Bhagirathi Kharak	1962–2001(39)	16.7
Chaurabari	1960–2010 (50)	6.5[a]
Pindari	1906-2001 (95)	15.2
Chipa	1961–2000 (94)	26.9
Meola	1912–2000 (88)	19.3
Jhulang	1962–2000 (38)	10.5
Nikarchu	1962–2002 (40)	9.2
Adikailash	1962–2002 (40)	12.8
Milam	1848–1997 (149)	16.7
Bhurpu	1966–1997 (31)	4.8
Mean	57 years	17.8

[a]Corrected from other sources (Source*:* Study Group Report, Government of India 2011)

Table 7.2 Inventory of major glaciers in river basins of CH (GSI 1999)

River basin	Number of glaciers	Glacierized area (sq. km)
Yamuna	52	143.4
Bhagirathi	238	755.4
Alaknanda	407	1255.1

is no precise method to estimate the period of its initiation and also to extrapolate or project retreat data for the future. It is also indicated that there are marked variations in data recorded at different places at different times suggesting fluctuating pattern of retreat which is inhomogeneous in space and time. Though the global warming is one of the potential reasons, it is difficult to pinpoint the total cause, thus making it more complex to separate local factors from the large-scale global changes and their impacts on Himalayan glaciers. Following tables (Tables 7.1 and 7.2) provide a glimpse of some relevant statistics and average retreat of the glaciers of CH.

It is a matter of general interest to introduce to the readers a few important glaciers of CH from among many glaciers out of which a major chunk remains yet unexplored. Gangotri, Pindari, Chorabari, and Dokriani are the choice of the authors based on the criteria of their significance and popularity. It may also be noted that each glacier is unique in its physical, esthetic, and morphological features; therefore, no generalization can be made based on exploration status and data of a few selected glaciers. One thing is sure that any adverse impact on the glaciers of the Himalaya will cause enormous hardship to many and will also create a sense of loss to a wide spectrum of society.

7.1.2.1 Gangotri Glacier

The Gangotri glacier is the one which has received the highest attention of glaciologists and common man alike since its first recorded observation in 1889 (Auden 1937), primarily due to its spiritual and religious importance. This glacier is one of the primary sources of water to the sacred river Ganga. The river Bhagirathi originating from the snout of Gangotri glacier is the most important tributary of river Ganga originating from one of the highest mountain peaks of CH. It has a length of about 30 km and its width varies from 2 to 4 km with many tributaries.

The Geological Survey of India maintains regular record of the snow position for Gangotri at regular intervals and their data show a retreat of approximately 19 m/year (based on data from 1935 to 1999) which is also supported by the recent satellite data monitored by Indian Space Research Organization. According to the worldwide glacier data compiled by NASA (USA), the Gangotri glacier has experienced recession at a rate of 19 m/year out of which the recession of last 25 years is reported to be at a much higher rate. Estimated recession rate of Gangotri glaciers from their records for the period 1996 to 1999 is stated to be about 25 m/year. Such picture marked with retreat lines is given by Fig. 7.1.

Fig. 7.1 Gangotri Glacier retreat lines drawn for the years 1780, 1935, 1956, 1964, 1971, and the year 2001 (free source image)

7.1.2.2 Chorabari Bamak Glacier

The Chorabari glacier in Garhwal Himalaya is about a kilometer from the Kedarnath temple. River Mandakini, a tributary of Alaknanda, originates from Chorabari and flows in south-west direction. The Kedarnath shrine is situated in the outwash of plains of Chorabari and companion glaciers within the flood plains of Mandakini river system. Quantitative record of the glacier position is available since 1962 onwards only. Studies through lichenometry and engravings on the walls of Kedarnath temple suggest that the glacier began its current retreat over two and a half century ago in 1748. The average estimated retreat of the glacier since 1960 is reported to be about 6.5 m/year. It is also inferred that the retreat has been faster and highly fluctuating during the last few decades (Chaujar 2012). Figure 7.2 provides a view of the Chorabari glacier with some visible land patches.

7.1.2.3 Pindari Glacier

The Pindari glacier is a glacier in the upper reaches of Kumaon Himalaya in district of Bageshwar to the peaks south-east of Nanda Devi and Nandakot ranges. Pindari is the most frequently visited glacier for adventure tourism though it is much smaller in area. This glacier has many records in the form of old photographs, glacier front

Fig. 7.2 Imagery of Aster satellite of NASA, showing Chorabari glacier and Kedarnath in Chamoli district of Garhwal Himalaya (free source image)

Fig. 7.3 A panoramic view of Pindari glacier in the Bageshwar district of Kumaon Himalaya (Courtesy: Anup Sah, FICS, Nainital)

location markings, and other evidences since 1845 within intermittent intervals. Average rate of glacier retreat based on last 100 years of data is reported to be 15.2 m/year. Figure 7.3 provides a glimpse of Pindari glacier with cumulus clouds over the mountain peaks.

7.1.2.4 Dokriani Glacier

Dokriani glacier (30°49′ to 30°52′N and 78°47′ to 78°51′E) is one of the well-developed, medium-sized glaciers of the Bhagirathi river basin in Garhwal Himalaya. It originates from Draupadi Ka Danda group of peaks at an elevation of 6000 m above mean sea level and is formed by two cirque glaciers. The glacier follows NNW direction for about 2 km before it turns towards WSW and terminates at an altitude of 3886 m. The length of the glacier is 5.5 km with a width varying from 0.08 to 2.5 km. The total catchment area is 15.7 km^2, with the glacier ice covering an area of 7.0 km^2. The thickness of glacier ice varies from 25 to 120 m between snout and accumulation zone; its average thickness is 50 m (Dobhal et al. 2004). A picture of Dokriani glacier is given in Fig. 7.4.

Fig. 7.4 Dokriani glacier in the Uttarkashi district of CH (Courtesy: Nilendu Singh, WIHG, Dehradun)

7.2 Climate Change and Water Resources

India in general is recognized as severe water stress region due to large natural variability of monsoons in time and space and also high rate of evaporation due to high temperatures throughout the year in most parts of the country. National Water Mission (MoWR 2008) has identified climate change and increased exploitation of all kind of water resources as the major threats to water sector in the country. To understand the impact of climate change on water resources, it is essential to examine the changes in complete hydrological cycle involving ocean, atmosphere, and the land in which the water exists in any of its three states or their combinations. The hydrological cycle is linked to the climate system through a chain of complex processes and feedback mechanisms; thus, an all-encompassing model may be required to make realistic quantitative impacts due to climate change. Reasonable statistical estimates can be worked out with available information.

Over the Indian region, the climate change is likely to adversely affect water balance due to changes in precipitation patterns, particularly the behavior of southwest monsoon, evapotranspiration, and rate of snowmelt. Increased frequency of droughts and floods may affect the groundwater and its quality in alluvial aquifers. High intensity rainfall associated with extreme weather events may lead to higher runoff particularly on mountain slopes which may result into faster water loss and consequent reduced groundwater recharge. In context of glaciers, the National Water Mission (NWM) document of the Ministry of Water Resources (MoWR 2008), Government of India have identified the expected decline in the glaciers and snow fields in the Himalaya as a potential threat to the perennial water resources. The Sutlej, Ganga and Brahmaputra basins the major water systems of south Asia are to be directly affected. According to this report, the major concerns of the

Himalayan region are the status of snow and glaciers, frequency of extreme weather events and large variability of flow across the rivers and channels with erosion, sedimentation and safety exceeding critical limits.

7.2.1 Climate Change and Water Resources: CH Specific Issues

The CH is endowed with favorable water supply position in terms of groundwater in the alluvial plains and snowmelt in the higher reaches of the mountains in addition to abundant rainwater during south-west monsoon and rain and snow in winter. The wide area of forest cover and lower temperatures during most part of the year are conducive for reduced water loss in evaporation and supportive to groundwater retention. In between the rainy or snowy season, sporadic rain spells associated with convective and orographic precipitation provide marginal precipitation to replenish the loss of water. It is therefore certain that any major change in the intensity and frequency of these weather patterns due to climate change are likely to alter the water supply position and thus affect the socioeconomic activity of the region and the adjoining areas downstream. These changes may significantly affect the water availability and its spatial and temporal variability and subsequently the water use pattern. Water availability in turn will affect agriculture, urban and rural house hold needs, sanitation, irrigation, livestock, fisheries and overall ecosystem balance including the forest cover.

Results of pilot studies conducted by many sponsored projects and official surveys for the high altitude rural regions of CH states point out that there is a general trend of depleting water availability at the source region. It is generally observed that the minor water sources such as freshwater springs, ponds, small and marginal lakes have been showing a sign of drying up during the course of last few decades in many mountain regions of the CH. These situations generally become acute during prolonged drought conditions which are not very uncommon and capable of producing prolonged water stress over a wider area. Local people invariably attribute these changes to modified rainfall patterns and erratic behavior of weather systems including the winter snow.

A visit to some of water sources and reservoirs during drought condition in CH will clearly demonstrate that the decline or disappearance of water sources at their origin is directly related to changes in land use pattern associated with population pressure and depletion of vegetation cover in and around watersheds. During the conditions of drought, scarcity of water may become further acute due to the prevailing unscientific and wasteful water management practices. Large-scale human interference at community to policy level makes it difficult to clearly estimate and delineate the impact of changing climate on water resource. Therefore, in case of water resources the human activities have a double pronged reaction, partly through the climate change and partly through short term unscientific resource management and distribution strategy in the event of major water scarcity.

Another major issue pertinent to the climate change impact on water resources in the region is the construction, maintenance, size and structure of major dams and minor reservoirs for irrigation and power generation. One of the important science questions which needs to be answered is how will the design parameters be decided for major water resource project in the background of projected climate change and the impact is assessed.

It is argued that the hydropower helps to reduce the greenhouse gases since it provides a non-emitting substitute to fossil fuel. Even if the role of major hydropower projects in affecting the global climate is assumed as insignificant, the likely impact of climate change on water resource and its management is very vital to be ignored.

7.3 Climate Change Forests and Biodiversity

Plant communities in general have direct linkages to the prevailing climate regimes over and around their habitat. Initial efforts in the documentation and classification of climatic zones over the globe have extensively and jointly used the available data on vegetation types and meteorological elements. With comprehensive data base on meteorological and biological elements being available now it is possible to study and model the sensitivities in interrelations between the two and develop bio-resource models to develop appropriate conservation strategies. Within the vegetated regions of the globe the forests occupy the largest part of the earth's surface in which the vegetation structure, distribution and ecology are greatly influenced by the climate. Carbon sequestration in forest ecosystem is the greatest sink of atmospheric carbon dioxide thus creating natural balancing mechanism to control the global warming. It is therefore logical to assume that changes in climate would alter the distribution characteristics of the forest ecosystem and conversely the forests will affect climate.

Forests and vegetation are the active ingredients of the global carbon cycle which can be viewed as a series of reservoirs of carbon in the earth system connected through carbon flux exchange mechanisms. The terrestrial biosphere reservoir contains carbon in all organic compounds in vegetation; living biomass, dead organic matter litter and the soils. CO_2 from the atmosphere is assimilated by plants through photosynthesis and the carbon fixed into plants is then cycled through plant tissues, litter and the soil carbon which can be released back into the atmosphere by microbial and respirational routes (Roger and Brent 2012).

There are many studies on laboratory experiments and computer simulation models suggesting that the increased CO_2 level in atmosphere will enhance the photosynthesis rate and thus increase the net primary product of forests increasing their carbon sequestration capacity. There are also studies suggesting that in a warmer world, the current carbon regulating service of forests in the form of carbon sink may get reduced substantially and may even become net source ultimately. The biospheric models using different climate change scenarios as suggested by IPCC are used to study the state and the pattern of change in forest cover over India. There are indications of shifts in forest types and the preliminary studies suggest an increase in net primary productivity as far as the Himalayan region is concerned (Ravindranath and Sukumar 1998; Ravindranath et al. 2006).

The mountainous regions of CH particularly the state of Uttarakhand is considered as a super biodiversity state of the country due to its geographical and climatic attributes. The availability of diverse life forms is more or less associated to normal health of the environment as rich biodiversity offers great resilience to the stresses posed by climate variability and change. Increased dependence of local population on bio-resources and their commercial exploitation further complicate the threat to biodiversity due to climate change. In some parts of high altitude mountain ranges of Himalaya, the biodiversity is being lost or endangered because of land degradation and overuse of resources, e.g., in 1995 about 10% of known species in Himalaya were listed as threatened (IPCC 2002).

A good understanding and comprehensive data on the climate impact and vulnerability profiles of the region would be helpful in identifying the most vulnerable regions and activities and production system including local practices to cope with the projected climate variability and change. In addition to the enhancement in the rate of extinction of endangered species in a forest ecosystem, the climate change may also induce migration and lateral movement of plant and animal species. In the Himalayan region, an upward shift of vegetation regimes due to increased temperature and receding glaciers is reported by many pilot studies and also reported by subjective observations of the local people. It is also reported from many high altitude stations that there are evidences of the alpine coniferous species of trees being replaced by low altitude broad leave species. These observations are supported by scientific studies suggesting changes in the composition of high altitude Himalayan forests, specifically the advancement of flowering, fruiting, and the maturity period in medicinal and herbal plants, weeds, and grasses and many endangered species.

7.4 Climate Change and Agriculture

It is well established that the global climate change will affect the entire agricultural production system through direct and indirect effects on crops, soil, livestock, pests, and diseases. There are studies suggesting that an increase in CO_2 concentration has a fertilization effect on crops through C3 photosynthetic pathways and thus promoting the growth and productivity. Most of these studies are based on field experiments using single agricultural crop for each study and under controlled environment. These inferences need much more extensive research to generalize the results. Many studies through biospheric model simulations or the field data on forests and biodiversity have varying conclusions when the role of nitrogen fertilization, net biomass productivity, grain yields, fruits, flowers, and biodiversity aspects are carefully examined in totality. In the meantime, it is also argued that the increase in temperature can reduce crop duration, increase respiration rates, and adversely affect photosynthetic process.

This may also influence survival and distribution of pests which would lead to new equilibrium between crops and pests and also alter the nutrient mineralization in soils and increase evapotranspiration. Studies on the role of biosphere as climate system component and its resilience to accept the change are currently at nascent stage and need vigorous research and development efforts though the sensitivity of agriculture to changing climate is well established. Mountain regions of the world produce a substantial portion of agricultural crops fruits and vegetables in terms of their economic value. In the upland region of Himalaya, the altitudinal gradients in climate lead to rapid changes in agricultural potential over comparatively short horizontal distances. The climate sensitivity of agriculture will vary according to the local geographical factors, types of crop, and the agricultural system and practices.

Farming in remote mountainous regions of the Himalaya has managed to strike a balance with fragile mountain environment and their agricultural practices for centuries. This balance is likely be disrupted by climate change and it would take a long time for new equilibrium state to be established. There are many constrains of mountain agriculture such as availability of cultivable land, soil type, water, landslides and soil erosion, climate extremes and access to profitable markets, transport and investment. In view of these constraints, any change in currently viable area for crop production and type of crops cultivated may not remain sustainable under the changed climate. In Uttarakhand, the agriculture is a highly complex integration of native and adapted practices on crops, animal husbandry, forest resources, and horticulture. In general, the response of agricultural crop production in different agro-ecological region to climate change varies according to crop composition, edaphic conditions, and the cropping pattern. The state has about 14% of its geographical area under cultivation with both rainfed and irrigated agriculture. Maximum productivity of agricultural crops is in the Himalayan foothill and plain areas; whereas, the productivity is much less in the high altitude mountain regions.

The agriculture in plains is characterized by commercial farming, mono cropping, agro-forestry, high irrigation, and use of fertilizers, pesticides, and mechanization; whereas, in the hills the agriculture is traditional, organic, and based on local ethnic crop varieties. Extremes of temperature and events of frost, snow, and dew have direct role to play in the winter crops in hills. In the mountains, the availability of rainwater and soil moisture storage is crucial to both winter and summer crops. Precipitation plays an important role in determining crop production in both rainfed and irrigated areas. In rainfed areas, sowing time, crop duration, and productivity are affected by quantity and distribution of rainfall; whereas, in the irrigated areas the germination and crop harvesting is usually affected by rainfall, its distribution in time and space. In addition, the source of irrigation is also a direct function of rainfall and its distribution in time and space; thus, the climate conditions are equally crucial to rainfed crops and the crops over irrigated lands.

Many agro-meteorologists have studied the crop–weather relationships for the crops of CH hills using rainfall and temperature data of last about three to four decades with many interesting results for crop management but no unilateral impact of climate change on agriculture seems to have been established yet. Most of the conclusions regarding the status of climate change and its impact on agriculture are based on the data collected at experimental farms of agricultural universities which is not ideally suited for conclusive climate change detection. It is also premature to arrive at the conclusive evidence of climate change impacts on agriculture due to large inter seasonal variability and fluctuating trends during different crop seasons depicted by relatively short periods of data (Yadav et al. 2014; Dubey et al. 2014). The crop-weather models are now being extensively used to study the climate change impacts on specific crops to develop climate resilient crops with basic assumption that the changing precipitation and temperatures are one of the major factors influencing crop growth and yield.

7.5 Climate Change and Human Health

Changes in environment at the global, regional, and local levels reflect changes in human health. The urban environment is deteriorating due to high density of population as urbanization has become an integral component of modern developmental goals. The detrimental impacts of environmental pollution on human health are profound and intimately related to climate variability and change and their precursors, the pollution. It is obvious that the changes in climate have linkages with accumulation, dispersal, and scavenging of pollutants of all kind in the air at local, regional, cross-national boundaries, and global level. The ultraviolet radiation reaching the earth's surface due to depleting stratospheric O_3 may cause many harmful effects on humans and flora and fauna.

Human health impacts of climate change are not yet systematically documented. Based on available data on communicable disease, their control, and spread along with the model simulation of climate change scenarios, it is possible to draw some broad conclusions on health impacts of climate change. Some conclusions drawn from such studies and reported in literature are briefly summarized for ready reference. These conclusions are derived from the studies conducted in many parts of the world based on limited data sets but are reasonable indicators for handling many health-related issues. It is assumed that the broad conclusions may also be relevant to the CH region in a broader sense with dominating local influences and factors. Some of them are listed below:

1. Number of deaths due to increased heat waves and lack of nighttime cooling would increase substantially in a warmer environment.
2. Infectious diseases like cholera would increase in some parts of the globe as emerging and reemerging infection takes root in the communities.
3. Risk of malaria transpiration would increase particularly in high altitude mountainous and mid-latitude regions. The malaria prone pockets along the foothills of CH can be taken as a pilot study area for different climate change scenarios.
4. Increase in humidity and temperature range with variable weather and higher frequency of extreme weather events would produce undesirable effect on human health and comfort.
5. Other mosquito-borne diseases like dengue and Japanese encephalitis are projected to become more prudent and spread dramatically as the climate changes to the warmer and wetter side.

6. Disease outburst due to enhancement in the frequency of ENSO events causing large anomalies in temperature and humidity may increase particularly in tropical regions.
7. Increased instances of floods and waterlogging and also the situations of droughts and famine affect the availability of freshwater and food making the populations more vulnerable to infection. The affected areas generate the sources of waterborne diseases and encourage the spread of communicable diseases coupled with resultant malnutrition.

7.6 Climate Change: Extreme Weather Events and Natural Disasters

Most of the natural disasters other than earthquakes, volcanoes, and tsunami are related to weather and climate anomalies. Many studies carried out for natural disasters using data from past observations and model simulations broadly suggest increase in weather and climate extremes and enhancement of natural disasters under the projected future climate change scenarios. There are suggestions that the instances of droughts, floods, heat and cold waves, landslides, avalanches, cyclonic storms, tornadoes, and the cloudbursts are all likely to increase in the projected warmer climate scenario. An increase in the frequency and magnitude of extreme weather and climate events due to climate change will affect the ecosystem degradation, reduced availability of water, food and fodder, thus impacting the livelihood. These impacts of climate change in consequence will reduce the capabilities of communities at large to cope with natural hazards. An example of devastation caused at Deoli village of Uttarakhand, due to heavy rainfall in 2010, is given in Fig. 7.5.

In India, there are many regions of potential vulnerability to natural disasters of high intensity short duration events to long period disastrous situations due to overall excess or scarcity of rains. Monsoon rainfall in India has large interannual variability characterized by sub-seasonal active and weak rainfall events. Heavy rain spells of longer durations; sometimes extending to a week or longer, generally induce landslides on geologically vulnerable lands and sloping terrains. As stated earlier during the break monsoon situation when the rainfall over the core monsoon region decreases, there is an increase in rainfall over Himalayan foothills (Pant and Rupa Kumar 1997). During most of the excess rainfall years in the country when monsoon is generally active and vigorous in large part of the country, the major part of CH also receives its share of rains and floods. Analysis of long period of available data on rainfall suggests that in general the heavy rain events show an increasing trend particularly in central India (Goswami et al. 2006; Rajeevan et al. 2008; Guhathakurta et al. 2011; Joseph et al. 2015; Revadekar et al. 2011).

There are many reports of extreme point rainfall (rain storms) from many parts of the state during the monsoon season as a result of continuous heavy rainfall over many days and also during pre- and post-monsoon seasons due to local convective storms invigorated by large-scale convergence. These events of heavy rain attain enhanced activity due to orographic lifting and strong convection at selected loca-

Fig. 7.5 Scene after disaster at Deoli village, Almora where ten people died due to heavy rainfall in 2010 (Courtesy: R Chandra, Nainital)

tion resulting in very heavy rain rates (precipitation intensity) referred to as cloudbursts. Cloudbursts cause flash flood, soil erosion, landslide, and damage to local ecosystem, landscape, life, and property. Though, it is still premature to relate the extreme weather events directly to climate change but many studies with model simulations and conceptual considerations point towards a link. It can be argued that in warmer atmosphere the concentration of net energy and instability of the atmosphere may increase under favorable conditions at preferred places to generate concentrated rain producing mechanisms.

In context of extreme weather events, the IPCC Working Group-1-Report (IPCC WG-1 2013, p. 122) states as follows. "Climate Change, whether driven by natural or human forcing can lead to change in the likelihood of occurrence and strength of extreme weather and climate events or both." Guhathakurta et al. (2011) have done an extensive study on extreme rainfall events based on the analysis of daily rainfall data for the period 1901–2005 using 2599 rain gauge stations including all stations which have more than 30 years data. Their study includes a few stations in CH and their analysis over the region shows an increasing trend in terms of 1-day extreme rainfall amount.

References

Ageta Y, Kadota T (1992) Predictions of changes of glacier mass-balance in the Nepal Himalaya and Tibetan Plateau: a case study of air temperature increases for three glaciers. Ann Glaciol 16:89–94

Auden JB (1937) Snout of the Gangotri glacier, Tehri Garhwal. Rev Geol Surv India 72:135–140

Chaujar R (2012) Climate change and its impact on the Himalayan glaciers—a case study of the Chorabari glacier, Garhwal Himalaya, India. Curr Sci 96(5):703

Dobhal DP, Gergan JT, Thayyen RJ (2004) Recession and morphogeometrical changes of Dokriani glacier (1962–1995), Garhwal Himalaya, India. Curr Sci 86(5):692–696

Dubey SK, Tripathi SK, Pranuthi G, Yadav R (2014) Impact of projected climate change on wheat varieties in Uttarakhand, India. J Agrometeorol 16(1):26–37

Goswami BN, Venugopal V, Sengupta D, Madhusoodanan MS, Xavier PK (2006) Increasing trend of extreme rain events over India in a warming environment. Science 314:1442–1445

GSI (1999) Inventory of Himalayan Glaciers: a contribution to the International Hydrological Programme. In: Kaul MK (ed) Geological survey of India, Special publication 34, p 165

Guhathakurta P, Sreejith OP, Menon PA (2011) Impact of climate change on extreme rainfall events and flood risk in India. J Earth Syst Sci 120(3):359–373

IPCC (2002) Climate change and biodiversity. Technical Paper V. Intergovernmental Panel on Climate Change. Assessment Report of the Intergovernmental Panel on Climate Change. WMO-UNEP

IPCC-WG-1 (2013) Climate change 2013. In: Stocker TF, Quin GK et al (eds) The physical science basis. Cambridge University Press, Cambridge. 1535p

Joseph S, Sahai AK, Sharmila S, Abhilash S, Borah N, Chattopadhyay R, Pillai PA, Rajeevan M, Kumar A (2015) North Indian heavy rainfall event during June 2013: diagnostics and extended range prediction. Clim Dyn 44:2049–2065. doi:10.1007/s00382-014-2291-5

Ministry of Environment Forests and Climate Change, Government of India (2011) Indian network for climate change assessment, climate change and India: a 4 × 4 assessment—a sector and regional analysis for 2030

MoWR (2008) Ministry of Water Resources River Development and Ganga Rejuvenation. Annual Report 2007–2008, p 99

Office of the Principal Scientific Advisor to the Government of India (2011) Report of the study group on Himalayan glaciers, 165p

Pant GB, Rupa Kumar K (1997) Climates of South Asia. John Wiley & Sons, New York/London. 320p

Rajeevan M, Bhate J, Jaswal AK (2008) Analysis of variability and trends of extreme rainfall events over India using 104 years of gridded daily rainfall data. Geophys Res Lett 35:L18707. doi:10.1029/2008GL035143

Ravindranath NH, Sukumar R (1998) Climate change and tropical forests in India. Clim Change 39:563–581

Ravindranath NH, Joshi NV, Sukumar R, Saxena A (2006) Impact of climate change on forests in India. Curr Sci 90(3):354–361

Revadekar JV, Hameed S, Collins D, Manton M, Sheikh M, Borgaonkar HP, Kothawale DR, Adnan M, Ahmed AU, Ashraf J, Baidya S, Islam N, Jayasinghearachchi D, Manzoor N, Premalal KHMS, Shreshta ML (2013) Impact of altitude and latitude on changes in temperature extremes over South Asia during 1971–2000. Int J Climatol 33:199–209. doi:10.1002/joc.3418

Revadekar JV, Patwardhan SK, Rupa Kumar K (2011) Characteristic features of precipitation extremes over India in the warming scenarios. Adv Meteorol 2011:1–11. doi:10.1155/2011/138425

Roger S, Brent S (2012) Carbon sequestration in forests and soils. Annu Rev Resour Econ 4:127–144

Yadav R, Tripathi SK, Pranuthi G, Dubey SK (2014) Trend analysis by Mann-Kendall test for precipitation and temperature for thirteen districts of Uttarakhand. J Agrometeorol 16(2):164–171

Chapter 8
Climate Change and Uttarakhand: Policy Perspective

8.1 General Background

Voluntary restrictions on the quantum of anthropogenic emissions of GHGs and mitigation and adaptation strategy for projected climate change require wide array of policy initiative and strategic actions at local, state, national, regional, and global levels. India will have to deal with the challenges of highly variable tropical monsoon climate, a highly fragile Himalayan ecosystem, large coastal belt, and the deserts.

G.B. Pant et al., *Climate Change in the Himalayas*,
DOI 10.1007/978-3-319-61654-4_8

High population density in vulnerable areas and rapid pace of development throws further challenge to policy makers and planners particularly at the state and local levels. It is thus imperative to take into account a scientifically evaluated adaption strategy along with appropriate mitigation measures and effective implementation of existing environment-related rules, regulations, and relevant laws at different levels.

It is obligatory for the policy makers that framing of new rules, regulations, and laws must cover concerned sectors on clean development mechanism safeguarding the interest of all stakeholders. Policy formulation will require a scientifically designated impact assessment mechanism using good quality high resolution quantitative data and models to provide necessary inputs. This exercise must be supported by appropriate networking of existing data to create knowledge base in the area of research, development, and awareness programs. The climate change vulnerability and impact assessment and also the mitigation measures must be supported by efficient use of emerging tools and technology for monitoring, analysis, and implementation.

The policies and programs relating to environment and climate in place and under operation at various levels of governance need to be periodically evaluated and updated. These initiatives at the national level will percolate down to state level and require the incorporation of state-specific issue in an overall policy document. In general, the climate change-related policies have their focus on promoting the understanding of climate change, adaption, mitigation, energy efficiency and natural resource conservation while pursuing overall sustainable economic development goals. A brief description of the policies relevant to the climate change issues specific to the state of Uttarakhand is presented for a few selected sectors of activities are as follows.

8.2 Agriculture and Allied Sectors

Overall agricultural practices and products in the hill and plain regions of Uttarakhand are different in many ways; therefore, these regions need to be dealt with separately as regards to climate change impacts on agricultural sector and development of climate resilient agriculture and adaptation strategies. Agricultural crops, practices, and marketing strategies are traditional for hills and to a large extent modern and commercially oriented in plain regions. Though, the farmer's traditional knowledge and expertise is the prime mover for all agriculture-related matters; inputs and guidelines of state agriculture, horticulture, fisheries, and animal husbandry departments are also taken into account.

Expertise of academic institutions such as ICAR-Indian Institute of Soil and Water Conservation (ICAR-IISWC), Dehradun, G.B. Pant Institute of Himalayan Research and Development, Almora, G.B. Pant University of Agriculture and Technology, Pantnagar, and other institutions under ICAR-IISWC and state agriculture department are prominent among them. Some of the important agricultural programs in the state will require implementation with the participation of local Non-Governmental Organizations (NGOs), stakeholders, Panchayats, progressive farmers, and private corporations. Role of government departments and private partners in the overall agricultural development have direct relevance to the climate change scenario for the state. These initiatives are at present negligible due to non-availability of an authentic framework on future climate change scenario as well as the scientific estimates of likely impacts at the state level and non-existent policy guidelines. In the absence of a clear picture of likely change in climate many uncertainties exist in deciding the priorities for the short- and long-term goals distinctly incorporating climate change component in agricultural planning.

Some of the programs which require impetus and refinements keeping in mind the impending consequences of climate change are watershed management, soil and water conservation, rainwater harvesting, agro-forestry, cultivation of medicinal plants, promotion of bamboo and fiber, organic farming, crop insurance, and plant protection. The guidance and advisory service related to crop planning and weather advisories need to be strengthened with the information network suitable for farmers. A brief introduction to some of these programs is given below.

8.2.1 Watershed Management

The watershed is the area of land bounded by a hydrologic system within which all living and non-living things are linked by their common water course. In a watershed, all of the water within it and which drains out of it is part of the same composite entity. Therefore, the fragile and precious water resource of this system needs to be meticulously managed. These programs need to be initiated with the help of all stakeholders within a scientifically designated watershed to achieve an integrated target of optimum resource use.

The main objectives of the watershed management program must be to improve the productivity potential of natural resources using modern conservation methods; specifically, soil, water, and nutrients as applicable to the local ecosystem as a sustainable integrated unit. Basic resource needs of the habitats and their future expansion and their impact on the ecosystem are required to be quantified and modeled for long-term planning. For the success of these programs, an active participation of local people and concerned agencies at the stages of conceptualization, planning, and implementation is most essential.

8.2.2 Soil and Water Conservation

Massive soil erosion in Uttarakhand is a concern particularly under changing climate scenario with projected increase in the occurrence of extreme weather events and frequent landslides. It is also to be kept in mind that the mountain regions of the state are located in high seismic activity zone thus requiring special attention. The landslide prone zones along the mountain slopes of Himalaya are in abundance primarily because of their basic geological features and top layers of loose soil and

Fig. 8.1 Erosion of Forest soil by Rispana Stream, Dehradun, Uttarakhand (Courtesy: Gita Pant)

debris. In conjunction with geology, the meteorological factors particularly the eroding and wash out influence of heavy rains playing crucial roles. An interesting example is given in the Fig. 8.1.

Water in the barren mountains with steep gradients is a great loss of precious resource during abrupt and heavy rains with rapid runoff. Water and soil conservation policies of the state need to be directed towards an integrated eco-friendly developmental strategy for road construction, urban infrastructure development, and irrigation schemes; particularly along the fragile hillslopes and riverbeds. High priority needs to be given to integrated afforestation and habitat conservation programs. State-of-the-art technology is to be applied in the activities of mining, transportation, landslide protection, drainage, and water supply schemes for irrigation and domestic use.

8.2.3 Input, Subsidy, and Support System

Timely availability of inputs for efficient agricultural activity such as seeds, fertilizers, irrigation, weather bulletins, and crop-related information is currently just at the sustenance level. An effective and viable policy on farm subsidy and crop insurance covering the risk due to natural hazards, explicitly including the climate change component is required. Due to its high level of vulnerability to climate change in Uttarakhand, new policies and their effective implementation is urgently required to cope up with increased climate risk, population increase, and an upsurge in demand and purchasing power of the masses. It is also very essential that the farm policies relating to production of quality seeds, timely availability of farm inputs, appropriate market for the products, and a comprehensive agricultural knowledge base must have climate change component in their basic fabric with scope for periodic upgradation and refinements.

8.2.4 *Organic Farming*

With increased awareness about the risk of health hazards in food products and general demand for food, poultry, fish, fruits, and vegetables with minimum contamination and proper nutrition balance, there is a global trend for increased interest in organic farm products. It is experienced that in general a substantial sector of the population may be willing to pay higher prices for these products as their awareness about the likely harmful effect of certain contaminants and resultant cost on medical remedies has become their primary concern. The organic farming policy is essential for improving crop productivity, soil health, and sustainable agricultural development. In addition to appropriate policy initiatives, the mass awareness and education on organic farming and allied subjects and authentic data on their health benefits and nutrition content are also to be provided to the consumers. Having assured an enhanced income to the farmers as an incentive the organic farm products will steadily establish their roots among the communities.

8.2.5 *Horticulture and Allied Agro-Horticultural Activities*

Horticulture is an important component of overall hill development. As an example, the success story of neighboring state of Himachal Pradesh is widely quoted and appreciated at many enterprises and official circles. The climate across the state right up to high altitude mountain regions is very favorable for wide range of fruits, vegetables, nuts, and flowers of extra tropical varieties. In addition, many wild varieties are grown in forests and reserve lands which have comparable nutrition content and commercial value specifically to the tribal people. In the changing climate scenario, many contingencies may arise due to shift in cultivation zones, impact of increased temperature and humidity at hill regions, and associated emergence of new pests and deceases. These eventualities may directly or indirectly impact the horticultural crops thus requiring scientific monitoring and guidance at different stages.

Traditional horticultural practices are established to cater for the requirements of water retention, reduction in soil erosion, generation of cattle feed and nutrition, and extra income to farmers. For a quantum jump towards net benefit and sustainable growth of these activities, an effective policy framework needs to be designed with priority. This can be achieved by creating an active interface of farmers with scientific and administrative groups within the concerned agricultural, horticultural, soil, water and land use planners, and other stakeholders at all levels including consumers. All the above-stated concerned groups should be fully equipped with the understanding and impacts of climate change at base level.

8.2.6 *Herbal, Medicinal, and Aromatic Plants*

During the last few decades, the cultivation of herbal, medicinal, and aromatic plants specific to the Himalayan region have received enhanced impetus. Many policies and guidelines are in place to provide appropriate information and guidance for cultivation, storage, transport, and in situ processing of these delicate commodities. Keeping in view the immense market and employment generation potential, the state government will have to encourage the establishment of new biochemical and biomedical industries and research and developmental centers to cater to the need of local products and the ecosystem.

Most of these plants and their products are highly climate sensitive and fragile and many among them belonging to the endangered species. It is therefore very essential that their conservation and application is covered under the mandatory policy guidelines. The policies and guidelines must have major component of research and incentive to conservation initiatives incorporating the component of future climate change scenarios. Ways and means must be found to realize their full potential and protect their natural habitat and avoid extinction of endangered species with appropriate measures for their growth and spread.

8.3 Forests and Biodiversity

Ecosystem, in general, experiences broad ranges of tolerance to environmental change and variability occurring in different timescales. It is very likely that they might have survived many episodes of minor climate fluctuations in the past. In a complex ecosystem, there will always be some species which are highly location specific and less tolerant to extremes and with delicate survival instinct. A stage in their life may arise when they may be pushed to the brink of likely extinction. The forest ecosystem in Himalaya is most vulnerable to climate change particularly to the species surviving at the critical margins of temperature and soil moisture stress due to extreme climate conditions.

Based on the projections of the future climate change scenarios, it is expected that the latitudinally upward and northward shift of species reported by many research workers may start appearing over a wider area in a relatively shorter time span. In addition, there are persistent phenological changes which may affect forest ecosystem. Other significant examples of adverse impacts on forest ecosystem may be that the long dry periods during summer may increase the frequency of forest fires adversely impacting the ecosystem. It is also likely that a prolonged warm and wet period may increase the instances of pests and deceases in agricultural crops and forests alike.

The forest department in Uttarakhand has a long tradition of planned exploitation, conservation, and protection of forest resources. A large number of policy guidelines are in place in the form of forest acts, rules, and regulations which have evolved with time focusing on afforestation, scientific planning, and management practices. These are designed and course corrected to cater to the growing need of natural resource growth and biodiversity protection. The state will have to be proactive in implementing these rules and regulations relating to forests, wildlife, and biodiversity reserves. An inbuilt flexibility component must be in place for all legal provisions to allow necessary future amendments keeping in view the changing needs arising out of the climate change. Biodiversity conservation is an extremely sensitive and delicate matter to be left solely to the officials and foresters. Education and participation of stakeholders of all kind is most essential. Management of biodiversity through motivated groups with sufficient participation of experts and local people of Panchayat, Block, and Tahsil/Taluka level are required. Involvement of people at grass root level to acquire adequate knowledge of biodiversity, species identification with local names and applications must be encouraged. Basic knowledge of conservation principles and practices imparted to local people will help in maintaining and sustaining biodiversity and may also generate economic benefits.

There are many steps which need to be taken urgently to preserve biodiversity and increase the resilience of endangered species to the impact of changing climate. Some of these are maintaining and restoring native ecosystem, scientifically managing the habitat of endangered species and documenting the available knowledge and to prepare local authentic biodiversity registers as a joint effort by experts, local people, commercial interest groups, and the governments.

It is also essential that a general awareness is created among the people about the necessity of maintaining and enhancing forest cover through afforestation, social forestry, agro-forestry, and mixed forestry with protected undergrowth. People in forests and peripheral areas should be taken into confidence to avoid extensive denudation, chopping, logging, grazing, forest fire, and protection of water bodies and wildlife in forests. Many other measures which will require long-term maintenance costs and investment will have to be implemented in phased manner by taking up the most climate-sensitive cases on priority basis. These may include the strengthening of network of ex situ (zoological parks, botanical gardens, seeds, and gene banks) and in situ (biosphere reserves, national parks, wildlife sanctuaries, lakes, and ponds) conservation sites. One of the vital components of mass awareness and knowledge development is the consolidation and documentation of traditional knowledge and expertise. As an important conservation practice, round-the-clock vigil on illegal trade, poaching and forest land intrusion is essential. These are essential to safeguard biodiversity, wildlife protection, and also to minimize the man–animal conflict in the forest areas.

8.4 Water and Allied Sectors

Water being the elixir of life on earth is perhaps the one which is most likely to witness the impact of climate change in a drastic manner. Water availability, conservation, distribution, and quality are all to be influenced in a multisource, multiuse, natural, and man-made management system. In major parts of the country, there is no optimum water use policy for user sectors such as the domestic, irrigation, industry, and hydropower generation. Guidelines, regulations, and enforceable laws are non-existent not only with regard to the quantum of water use but also the quality standards and conservation practices. Following are some of the specific areas of water sector which require urgent attention keeping in mind the likely impacts of climate change directly to the water sector and their repercussion on allied sectors and their vital life-threatening consequences if not addressed with due urgency and care.

8.4.1 Water Quality, Safety, and Sanitation

There is an urgent need for guidelines and legislations with strict enforcement strategy of water quality, water supply, and wastewater management. In villages and public places, water is used from multiple sources such as rivers, springs, canals, and ponds which are generally overexploited and polluted, thus posing serious threat to human life including the very survival of humans and domestic and wild animals. These activities disturb the ecosystem and also increase the burden of health care cost and reduce working efficiency of the populations. In case of subsurface water sources, the quality and quantity of downstream flow is linked to the inputs upstream and near the source region. In the eventuality of abrupt climate change, there is a possibility of serious impacts on water resources and in all probability, there is a likelihood that it may result into serious health and social unrest problem because of contaminated and depleted water resources.

8.4.2 Community Preparedness and Partnership

"Conserve and keep the water clean" must be the philosophy not only in urban areas but also in the rural areas along with an intensive awareness program for rainwater harvesting and wastewater recycling. Programs of scientific watershed management and conservation practice must be vigorously pursued all over the state and supply of clean water must be linked to community health programs. These programs must have an active participation of state governments, local self-government organizations, social, religious, and non-governmental organizations.

The concerned parties must be fully equipped with easy-to-use practices and scientific monitoring including inbuilt training and awareness components. It is surprising to note that the Uttarakhand being the major source region of freshwater there are hardly any institutions specifically to create human resource (technicians, managers, and educators) for the water sector in its totality. In contrast, there are many research institutes, universities, and organizations devoted to research and training in agriculture, horticulture, forestry, petroleum, power, and such other sectors. Water as a natural resource like air does not recognize political or administrative boundaries; therefore, water resource and utilization issues will have interstate components to be resolved in a larger context.

8.4.3 Water Use Pricing and Regulations

In view of expected stress on water resources due to changing climate over the source regions, it is necessary that a water use policy document is available with respect to current status and future considerations of climate change, urbanization, and population increase. The framework of this policy should be based on the fact that providing safe and clean water to all for their indoor and outdoor applications is one of the primary responsibilities of the state. The water is essential, precious, and reasonably chargeable commodity for the users at par with food and shelter. State policy on water use must have an adequate component of recovery of cost from the users under a transparent and reasonable cost-price policy. The water resource, distribution, and use policy must incorporate due consideration for establishment of scientific planning, modern technology, and proper prioritization of resources as a component of a responsible welfare governance agenda.

8.4.4 Water Conservation Policy and Programs

Most of the water sources in the state of Uttarakhand can be considered perennial with replenishment by rains during monsoon, winter snow, rain showers in other seasons, as well as the snowmelt from glaciers and mountains. Due to steep slopes and rocky terrain, the groundwater storage is very limited in hills but is a highly significant component in the foothills of mountains and the plain regions. The large space-time variability of rainfall over the state results into uncertain rainwater availability for major rivers and their tributaries. To cope up with the eventuality of long dry spells or extended periods of droughts in the state, it is highly desirable that during the monsoon season rainwater is stored in ponds, natural water bodies together with use of scientific measures for water harvesting, conservation, and enhancement in the capacity of groundwater recharge.

These measures require joint and collaborative efforts from concerned government departments, user agencies, and community at large. It is suggested that the state agencies and departments such as forest, agriculture, irrigation, soil conservation, municipal boards, and power companies jointly formulate a common minimum program on short- and long-term basis for the conservation of water resources and their optimum use.

8.4.5 Induction of Environment-Friendly Optimum Water Use Technology

In traditional methods of domestic, industrial, and farm use of water minimum initial cost and abundant water availability have been the criteria for adaptation of various practices and technologies. With emerging scenario due to climate change and pressures of population increase, urbanization, and overall infrastructural development, it has become necessary that new and optimized water use habits are inculcated among the masses and they are also provided incentives for efficient methods and new technology. These may include new optimum use sanitary devices, sprinklers, and drip irrigation techniques to reduce losses due to evaporation and overuse as well as a scientific and engineering strategy to combat water losses in transit and evaporation from open ponds, canals, and reservoirs. Use of recycled water can be made mandatory for certain non-domestic applications such as irrigation, gardening outdoor cleaning. Water quality testing and monitoring under scientifically laid down standards and a reasonable price tag will create a sense of value to water and its importance to all individual and group users.

8.5 Natural Disasters: Reduction and Relief

The frequency and spatial distribution of natural disasters, commonly associated with extreme weather events, are likely to increase with warmer climate as a result of overall additional heat energy and moisture induction into the atmosphere. The weather systems over the Himalaya may become highly sensitive to the enhancement of energy inputs to the local intense weather systems. Flash floods, droughts, forest fires, landslides, cloudbursts, cold and heat waves, and avalanches are the phenomena in Himalayan states which quite often culminate into events of disastrous proportions. In recent years, there are many instances of high intensity rainfall over smaller regions occurring in short spans of time, generally associated with cloudbursts and intense convective storms.

In Uttarakhand, the cloudburst of Tehri on 31 August 2001, Budhakedar on 10 August 2002, flash flood and landslide of Bhagirathi valley in June 2000, and heavy rains with cloudbursts of 16–17 June 2013 in Kedarnath are some of the disastrous examples. Considering the difficulties of poor infrastructure and logistics coupled with short lead time and uncertainties of weather forecast, the relief and rescue operations require highly integrated and coordinated effort on a multi-pronged manner. On the lines of the National Disaster Management Authority (NDMA), the state must have a disaster warning, monitoring, and relief mechanism. The mechanism may primarily involve the India Meteorological Department, Indian Space Research Organization, State revenue department, Voluntary organization, and Police and Paramilitary forces. Many institutions in the state are actively involved in use of GIS, GPS, and satellite data to prepare inventory of these events and develop early warning system. These information and techniques must be used through a carefully crafted policy and mechanism for exchange and sharing of knowledge, data, experience, and resources for prediction, prevention, and relief operations.

8.6 Industries and Transport

Some of the important steps required to be initiated by the state administration as a part of their major industrial policy taking into account climate change factor are briefly mentioned as follows. The mining of limestones in mountains and valleys as well as transportation of sand and stones from riverbeds can pose great danger to the fragile ecosystem and may trigger landslides, soil erosion, atmospheric pollution, and alterations in channel flows. These activities may cause substantial damage in the event of increase in weather extremes as witnessed the recent years (Kedarnath episode 2013). These activities need scientific planning and proper guidelines to impose necessary restrictions with provision of regular scrutiny and appropriate penalty for violations. According to existing policy norms, the activities involving horticulture, floriculture, medicinal and aromatic herbs and plants, honey, agro-industries, silk, wool, woven natural fabrics, sports goods, paper products, pharmaceutical products, information technology, ecotourism, handicraft, woodwork, and local material-based packaging industries are a few of the thrust sector industries. These industries are eligible for special incentive and encouragement from environmental point of view.

However, in case of other heavy and medium industries with potential for water, air, and environmental pollution, the incentives may be available only if these are located within the designated industrial areas or parks. In addition, they should also generate higher level of employment and follow the norms of air and water quality standards laid down from time to time. These steps at present appear to be more of regulatory nature and lack desired infrastructure for scientific monitoring and evaluation. It is also necessary that the mandatory guidelines are periodically updated with new data on climate and environment-related parameters with strict compliance norms. Specifically, for the industrial development in high altitude mountain regions a cautious approach supported with sufficient data and mandatory environment impact analysis is necessary. An attempt in this direction will have to be made by the state and local administration to develop an appropriate industrial development policy for the regions. The document may contain a long-term plan for the thrust sector industries and their most suitable locations. These guidelines may have to be essentially based on the premises that the industrial development should be restricted to eco-friendly and non-polluting types depending on the locally sustainable inputs.

In respect of transport sector, some of the steps to mitigate the impact of climate change include the changing over to fuel efficient vehicles with well-implemented emission standards, use of hybrid vehicles, use of biofuels, and possible change from road to rail transport in plain regions. Development of helipads and medium airports at appropriate locations with meticulously planned high altitude aviation strategy for tourism, defense, and disaster mitigation objectives is likely to reduce GHG emissions. It is also necessary that the road transport to difficult terrains and economically non-viable locations is developed with good quality mountain trails and walk paths in the range of say less than 5 km at all safe and scenic locations.

The manufacturer's authorised representative in the EU is Springer Nature Customer Service Centre GmbH, Europaplatz 3, 69115 Heidelberg, Germany. If you have any concerns regarding our products, please contact ProductSafety@springernature.com

Printed and bound by CPI Group (UK) Ltd, Croydon, CR0 4YY
30/09/2026
02225577-0008